Archaeology in Environme and Technology

T0179028

"This important volume examines the indivisibility of environment, technology and human society in studies which cover the field of archaeology from the palaeolithic to the historic, from artefact studies to cultural resource management, and from the Pacific to the Mediterranean."
—*Harry Allen, University of Auckland, New Zealand*

Environments, landscapes and ecological systems are often seen as fundamental by archaeologists, but how they relate to society is understood in very different ways. The chapters in this book take environment, culture and technology together. All have been the focus of much attention; often one or another has been seen as the starting point for analysis, but this volume argues that it is the study of the inter-relationships between these three factors that offers a way forward. The contributions to this book pick up different strands within the tangled web of intersections between environment, technology and society, providing a series of case studies which explore facets of this common theme in different settings and circumstances and from different perspectives. As well as addressing themes of theoretical and methodological interest, these case studies draw on primary research dealing with time periods from the late Pleistocene glacial maximum to the very recent past, and involve societies of very different types. Running through all the contributions, however, is a concern with the archaeological record and the ways in which scales of observation and availability of evidence affect the development of questions and explanations.

The diversity of the chapters in this volume demonstrates the inherent weakness in any attempt to prioritise environment, technology or society. These three factors are all embedded in any human activity, as change in one will result in change in the others: social and technical changes alter relations with the environment—and indeed the environment itself—and as environmental change drives changes in society and technology. As this book shows, it is possible to consider the relationship between the three factors from different perspectives, but any attempt to consider one or even two in isolation will mean that valuable insights will be missed.

David Frankel is Professor of Archaeology in the School of Historical and European Studies at La Trobe University, Australia.

Susan Lawrence is Associate Professor of Historical Archaeology in the School of Historical and European Studies at La Trobe University, Australia.

Jennifer M. Webb is a Charles La Trobe Research Fellow in the School of Historical and European Studies at La Trobe University, Australia.

Routledge Studies in Archaeology

1 **An Archaeology of Materials**
Substantial Transformations in
Early Prehistoric Europe
Chantal Conneller

2 **Roman Urban Street Networks**
Streets and the Organization of
Space in Four Cities
Alan Kaiser

3 **Tracing Prehistoric Social
Networks through Technology**
A Diachronic Perspective on the
Aegean
Edited by Ann Brysbaert

4 **Hadrian's Wall and the End of
Empire**
The Roman Frontier in the 4th
and 5th Centuries
Rob Collins

5 **U.S. Cultural Diplomacy and
Archaeology**
Soft Power, Hard Heritage
Christina Luke and Morag M. Kersel

6 **The Prehistory of Iberia**
Debating Early Social Stratification
and the State
*Edited by Maria Cruz Berrocal,
Leonardo García Sanjuán and
Antonio Gilman*

7 **Materiality and Consumption in
the Bronze Age Mediterranean**
Louise Steel

8 **Archaeology in Environment
and Technology**
Intersections and Transformations
*Edited by David Frankel, Jennifer
M. Webb and Susan Lawrence*

Archaeology in Environment and Technology

Intersections and Transformations

**Edited by David Frankel,
Jennifer M. Webb
and Susan Lawrence**

LONDON AND NEW YORK

First published 2013
by Routledge

2 Park Square, Milton Park, Abingdon, Oxfordshire OX14 4RN
52 Vanderbilt Avenue, New York, NY 10017

*Routledge is an imprint of the Taylor & Francis Group,
an informa business*

First issued in paperback 2019

Library of Congress Cataloging-in-Publication Data

Archaeology in environment and technology intersections and transformations /
    edited by David Frankel, Jennifer M. Webb and Susan Lawrence.
        pages cm. — (Routledge studies in archaeology ; 8)
    Includes bibliographical references and index.
    1. Environmental archaeology—Case studies.   2. Landscape
archaeology—Case studies.   3. Human geography—Case studies.
4. Agriculture, Prehistoric—Case studies.   5. Tools, Prehistoric—Case
studies.   I. Frankel, David, 1946– author, editor of compilation.
II. Webb, Jennifer M. autor, editor of compilation.   III. Lawrence,
Susan, 1966– author, editor of compilation.
    CC81.A67 2013
    930.1—dc23
    2012045042

ISBN: 978-0-415-83201-4 (hbk)
ISBN: 978-0-367-86816-1 (pbk)

Typeset in Sabon
by Apex CoVantage, LLC

Contents

List of Figures   vii
List of Tables   ix

1   Complex Relations: Intersections in Time and Space   1
    DAVID FRANKEL, JENNIFER M. WEBB AND SUSAN LAWRENCE

PART I
Responses to Environments

2   Perspectives on Global Comparative Hunter–Gatherer
    Archaeology: Glacial Southwest Tasmania and Southwest
    France   13
    RICHARD COSGROVE, JEAN-MICHEL GENESTE, JEAN-PIERRE CHADELLE AND
    JEAN-CHRISTOPHE CASTEL

3   Strategies for Investigating Human Responses to Changes
    in Landscape and Climate at Lake Mungo in the Willandra
    Lakes, Southeast Australia   31
    NICOLA STERN, JACQUELINE TUMNEY, KATHRYN E. FITZSIMMONS AND
    PAUL KAJEWSKI

4   A New Ecological Framework for Understanding
    Human–Environment Interactions in Arid Australia   51
    SIMON J. HOLDAWAY, MATTHEW J. DOUGLASS AND PATRICIA C. FANNING

5   Integrating Hunter–Gatherer Sites, Environments, Technology
    and Art in Western Victoria   69
    DAVID FRANKEL AND CAROLINE BIRD

6   Pushing the Boundaries: Imperial Responses to Environmental
    Constraints in Early Islamic Afghanistan   84
    DAVID C. THOMAS

PART II
Technology and the Environment

7 A Long-Term History of Horticultural Innovation and
   Introduction in the Highlands of Papua New Guinea   101
   TIM DENHAM

8 Agricultural Economies and Pyrotechnologies in Bronze
   Age Jordan and Cyprus   123
   STEVEN E. FALCONER AND PATRICIA L. FALL

9 Changing Technological and Social Environments in the
   Second Half of the Third Millennium BC in Cyprus   135
   JENNIFER M. WEBB

10 Landscape Learning in Colonial Australia: Technologies
   of Water Management on the Central Highlands Goldfields
   of Victoria   149
   SUSAN LAWRENCE AND PETER DAVIES

PART III
Nature and Culture

11 Exploring Human–Plant Entanglements: The Case of
   Australian *Dioscorea* Yams   167
   JENNIFER ATCHISON & LESLEY HEAD

12 People and Their Environments: Do Cultural and Natural
   Values Intersect in the Cultural Landscapes on the World
   Heritage List?   181
   ANITA SMITH

   Contributors   205
   Index   211

Figures

2.1 Location of Combe Saunière, southwest France 16
2.2 The stratigraphy of Combe Saunière 17
2.3 Plan of Combe Saunière showing squares H16–H21 18
2.4 Plotted artefacts and fauna at Combe Saunière 20
3.1 Map of Australia showing the main river systems draining the southeast Australian highlands and the location of the Willandra Lakes 32
3.2 Geological map of the Willandra Lakes region 34
3.3 Map of part of the northern Mungo lunette 38
4.1 An aerial image of the Rutherfords Creek and Howells Creek catchments draining into Peery Lake in western New South Wales 63
5.1 Map of Gariwerd showing the location of sites mentioned in the text 70
5.2 Summary of the chronological framework for assemblages and rock art phases 71
5.3 Proportions of main raw material types in each assemblage 73
5.4 Schematic models of land use patterns in and beside the ranges 81
6.1 The Ghūrids' sphere of influence at its maximum extent ca 1200 CE 87
6.2 Minaret of Djām and surrounding mountain slopes, pock-marked by robber holes 88
6.3 Estimated extent of the Ghūrid summer capital at Djām 89
7.1 Map of Papua New Guinea with inset, showing sites mentioned in the text 104
7.2 Chronology of practices and forms of plant exploitation in the Upper Wahgi Valley 105
7.3 Composite image showing practices discussed in the text 106

7.4 Base of former mounds dating to 6950/6440 cal BP at Kuk 112

7.5 Multiple ditches exposed in the base of an excavation at Kuk 113

8.1 Map locating the Bronze Age settlements of Tell el-Hayyat and Tell Abu en-Niʻaj, Jordan, and Politiko-*Troullia*, Cyprus 125

9.1 Map of Cyprus showing the location of excavated Late Chalcolithic and Philia EC settlement sites and tombs 137

9.2 a. Relative proportions of five major species in faunal assemblages from Chalcolithic Lemba and Kissonerga and Philia EC Marki; b. Relative proportions of rubbers and querns in the curated ground stone assemblages from Chalcolithic Lemba and Kissonerga and Philia EC Marki 139

9.3 Predicted range of behavioural variability at times of high and low adaptedness in Late Chalcolithic and Philia EC communities over time 141

9.4 Occurrence of grave types at Kissonerga during the Middle (Period 3) and Late Chalcolithic (Period 4) 142

10.1 Plan of central Victoria showing main creeks and rivers, and goldfields 152

10.2 Annual rainfall recorded at Ballarat Survey Office and Bendigo Prison from 1858 to 1889 153

10.3 Small holding dam at Humbug Hill, near Creswick in central Victoria 158

10.4 Water race to Humbug Hill, near Creswick in central Victoria 159

11.1 Excavated *Dioscorea transversa* tuber *in situ* 170

11.2 Digging for *Dioscorea transversa*, Keep River, Northern Territory 173

11.3 Yam tuber (*Dioscorea transversa*) dug from loose sandy soil, Keep River region, Northern Territory 174

11.4 Recently excavated yam hole on small rocky dolerite outcrop, Keep River region, Northern Territory 175

Tables

2.1 The age range for each cultural phase and corresponding
time span is shown, as determined from the dating at
Combe Saunière   19

3.1 Archaeological traces, their stratigraphic, sedimentary
and palaeoenvironmental context and approximate age   39

3.2 Palaeoenvironmental context and approximate age
of each artefact assemblage   44

5.1a Summary data on stone assemblages: Early assemblages   74

5.1b Summary data on stone assemblages: Late and
Final assemblages   75

5.2 Summary of climatic changes over the last 16,000 years
in southwest Victoria, based on modelling of the
precipitation/evaporation ratio   79

6.1 Population estimates for Ḏjām based on population per
hectare, derived from modern ethnographic studies in Iran   90

8.1 Bronze Age chronologies for the southern Levant and
Cyprus, showing general chronological relationships
between Tell el-Hayyat, Tell Abu en-Ni'aj and
Politiko-*Troullia*   124

8.2 Relative frequencies of identified bone fragments by animal
taxon excavated from Tell Abu en-Ni'aj (Phases 6–1), Tell
el-Hayyat (Phases 5–2) and Politiko-*Troullia* (latest phase in
Troullia East and West)   128

8.3 Relative frequencies of identified seeds, seed and charcoal
densities and seed:charcoal ratios for Tell Abu en-Ni'aj
(Phases 6–1), Tell el-Hayyat (Phases 5–2) and Politiko-*Troullia*
(latest phase in *Troullia* East and West)   129

12.1 World Heritage Cultural Landscape Categories   185

12.2 Cultural landscapes inscribed on the World Heritage
List in 2012   186

12.3 World Heritage cultural landscapes associated with
'hunter–gatherers' or traditional (non-European) societies   197

1 Complex Relations

Intersections in Time and Space

*David Frankel, Jennifer M. Webb
and Susan Lawrence*

In his classic study *Habitat, Economy and Society*, Forde addressed the general relationships between these three factors in a wide variety of ethnographically described peoples with different social forms and subsistence technologies in many parts of the world, making "it possible to appreciate the complex relations between the human habitat and the manifold technical and social devices developed for its exploitation" (Forde 1934: 460). His focus on descriptions of particular groups and their lifeways was deliberately intended to counterbalance tendencies toward broad generalisations where "the reality of human activity escapes through so coarse a mesh" (1934: 460). Although Forde recognised that the "economic and social activities of any community are the products of long and intricate processes of cultural accumulation and integration, extending back in time . . ." (1934: 465), specific historical developments lay beyond his purview. This remains the task of archaeologists and historians. In this book some of these 'complex relations' are considered primarily from an archaeological perspective, providing the depth of time needed to consider how relations evolved over decades, centuries and millennia.

Environments, landscapes and ecological systems are often seen as fundamental by archaeologists, but how they relate to society is understood in very different ways. One approach is the privileging of physical context, as in various forms of 'possibilism' where the environment is given explanatory power as it provides constraints or opportunities for different historical, social or economic patterns (Ellen 1982) or more subtly influences settlement and relationships. So, too, we find the groundbreaking work of Cyril Fox, whose *Personality of Britain* (1932) saw the environment as an influential setting (but not a determining structure) within which historical events were enacted, taking a humanistic view of the interplay between cultures and contemporary conditions. While Fox placed greater emphasis on the environment as a shaper of society, the participants in the 1955 Wenner-Gren symposium on *Man's Role in Shaping the Face of the Earth* (Thomas & Sauer 1956) turned their attention to the other side of the environment/society equation. Symposium organiser Carl Sauer's influential concept of 'cultural landscape' explicitly drew attention to human action in shaping landscape, which he saw as the product of people acting on the natural world (Sauer 1925: 46; Anschuetz et al. 2001: 163). In the UK similar views informed the work of W.G. Hoskins, whose book *The Making of the English Landscape* (1965) took what Ashmore (2004: 258) has called a 'genealogical' approach to explaining the formation of traditional landscapes and set the agenda for generations of British scholarship.

These parallel traditions of environment-oriented and socially oriented research have continued and have been equally important. The recognition of environmental context was a major factor in the development of economic archaeology (Clark 1957) and was largely responsible for the expansion of specialist studies, often grouped under the broad umbrella of Environmental Archaeology (Dincauze 2000). Despite its importance, this is all too often divorced from 'mainstream' archaeology (especially where seen as the domain of specialist 'environmental scientists'), despite calls for closer integration (Dincauze 2000: 498, 512). The question is how an appropriate integration of environmental data and broader archaeological agendas may be achieved. In part this depends on who is asking the question, and whether the aim is, for example, 'anthropocentric palaeoecology' (Dincauze 2000: 498) or a more socially nuanced analysis, such as that of Evans, who looks to the development of theory to link social constructs and *mentalité* to people's perception of place and context (2003: 250).

While environmental archaeology is underpinned by theory from the natural sciences, social theory of various kinds has contributed to the development of cognitive approaches to landscape that emphasise meaning, memory and people's attachment to place (Ashmore 2004; Deagan 2008; Hardesty & Fowler 2001; Holtorf & Williams 2006; Mrozowski 2006). Social approaches have expanded enquiry beyond economic and functional aspects of the environment to consider questions such as how power and ideology may be embodied in the landscape (e.g. Bender 1993; Leone 1984) and how this may be of relevance to modern communities where the meaning of landscape is entangled with issues of access and control (Knapp & Ashmore 1999). Broader social and environmental concerns have also led to renewed interest in the unexpected, unwitting and often indirect impacts of people on the environment and their consequences, both on a global and more local scale (Kirch 2005).

Considerations of technology offer a means of opening up the traditional interplay between environment and culture. While the environment together with cognition and symbolic meanings of artefacts may be regarded as keys to exploring social relations (Evans 2003: 250), the more mundane, practical aspects of artefacts and their use is of equal importance. Technology, in its broadest sense, has always been understood to provide an important interface between people and the environment, the means to make use of resources of all kinds in many different ways. Technologies may include a broad understanding of how to use land and resources as well as material equipment; and while these technologies of action and equipment are always deeply socially embedded, affected by and affecting behaviour in many ways, not all lead to the best outcomes in either the short or longer term for people or their environment (Lemonnier 1993). As with other factors, technology can be approached from several perspectives, which give priority variously to practical, conceptual or social dimensions (Miller 2007: 3–7). These may include the specific engagement with resources (as, for example, studies of 'ceramic ecology' (Arnold 1993)) or, on a broader scale, the major issue of the development of agriculture, involving as it does substantial differences in all facets of behaviour and connection to the physical world. Introductions of new techniques (and associated concepts and equipment) may often be more difficult than expected (Guille-Escuret 1993) or have unexpected but significant social outcomes (such as those documented in Salisbury's (1962) study of the consequences of technological introductions in Papua New Guinea). From an archaeological perspective technological evidence may also serve as a proxy measure (often the only measure) of how past peoples operated within their worlds: technology here becomes an archaeological tool as much as a part of explanation.

Here we take environment, culture and technology together as our main axes of interest. All have been the focus of much attention; often one or another has been seen as the starting point for analysis. We argue that whatever the starting point, it is the study of the interrelationships between these three factors that offers a way forward. The contributions to this book reflect some of the inherent diversity of approach, and explorations of the interplay between the three axes are conditioned as much by context and material as by theoretical constructs. These studies also highlight the importance of different scales of time and space and rates of change in climate and environment compared with those of cultural processes, and how we choose our explanatory models and integrate diverse forms of evidence into our arguments. We cannot use the same mode of explanation for the short-term, immediate impacts of specific historical events or extreme natural events as we can for technological innovations or long-term, perhaps very slow natural developments.

Most of the chapters in this book are based on research discussed at a workshop at La Trobe University in Melbourne in 2010. They pick up different strands within the tangled web of intersections between environment, technology and society, providing a series of case studies which explore facets of this common theme in different settings and circumstances and from different perspectives. Although the majority of the contributors are based in Australasia and half the chapters deal with Australian hunter–gatherers, this Antipodean standpoint is, to our minds, no bad thing, providing as it does alternatives to more familiar northern-hemisphere perspectives. The question of adaptations to different conditions has long been a feature of Australian research (e.g. Mulvaney & Golson 1971;

Dodson 1992; Allen & O'Connell 1995), especially as Australia and Papua New Guinea provide unique access to long-term occupation in some of the world's most challenging environments and evidence of human adaptation in one place over millennia of environmental change. They also provide access to data on adaptations and innovation during a very recent colonisation event, the settlement of non-Aboriginal people in the region from 1788 onwards. Alongside the archaeological and environmental evidence for this colonisation, researchers can draw on documentary data to explore some of the social and cognitive dimensions of human–environment–technology relationships that are beyond the reach of many archaeological studies.

In Australia working with indigenous communities has developed awareness of the complexity and richness of traditional perceptions of land and resources as active and significant forces; but the variable and contingent elements of the Dreaming provide, at the same time, a recognition that specific knowledge of these levels of understanding cannot be readily or glibly incorporated into archaeological understanding. This is especially so where the long timescales and extensive spatial scales endemic to Australian research are taken into account. The vast areas of territories and the fluctuations in resource reliability and richness over tens of thousands of years of major climatic fluctuations have an inevitable effect on research, spilling over from local prehistory into studies of other times and places. The need to integrate specifics of artefact and site into these larger structures militates against the uncritical projection of current theoretical concerns onto complex archaeological data.

As well as addressing themes of theoretical and methodological interest, our case studies draw on primary research dealing with time periods from the late Pleistocene glacial maximum to the very recent past, and involve societies of very different types. This diversity flows through to considering varied scales of time and space. Comparisons through time or of different social or technological systems are seen both within individual contributions as well as between them.

We have chosen to group the chapters into three sections, although there is no simple separation between them. In some cases it is a matter of emphasis and focus rather than substance. In the first section the environment in all its complexity emerges as the point of departure, influencing social and cultural patterns: the archaeological analysis of technological systems provides a means for monitoring adaptations and adjustments. This is most obvious in the four studies of hunter–gatherers adjusting to varied climatic and ecological conditions; but the difficulties of maintaining a sustainable social system in an unforgiving context can also be seen in more complex or larger-scale societies. The second section groups studies where technological choices and systems come to the fore, either defining alternative uses of similar environments or as one means employed to respond to or overcome environmental conditions. The two chapters in the third section expose deeper issues regarding social or academic definitions of nature. Running through all the contributions, however, is a concern with the archaeological record and the ways in which scales of observation and availability of evidence affect the development of questions and explanations.

Questions of resolution and comparability of data are a particular feature of several chapters in the first section on 'Responses to environments.' Here the environment provides the starting point for explorations of technological and social adaptation. Richard Cosgrove, Jean-Michel Geneste, Jean-Pierre Chadelle and Jean-Christophe Castel are specifically concerned with establishing a common approach and common measures that will allow meaningful comparisons between Late Pleistocene hunter–gatherers in different parts of the world. Without this, they argue, it is not possible to develop broader archaeological models of hunter–gatherer behaviour. The protocols they are developing for analysing faunal and artefactual assemblages from Tasmania and France should provide a more secure base for assessing the nature of social and technological responses to extreme climatic conditions in these two archaeologically rich areas. The variable archaeological records imply equivalent variable distributions and densities of human populations and resources across landscapes during glacial and interglacial cycles, encouraging questions about how climate affects cultural trajectories and shifts in the intensity of occupation at a global scale.

Similar issues of disentangling diverse sources of evidence are also seen in Nicola Stern, Jacqueline Tumney, Kathryn Fitzsimmons and Paul Kajewski's investigation of human responses to landscape and climate change in semi-arid western New South Wales. Here the specific difficulties of working in this region impose constraints on what can be done, and require—as does most research—the development of methods tailored to the particular situation. The complex archaeological record and evolving contextual setting nevertheless provide the basis for assessing the long-term history of human settlement in relation to changing palaeoenvironments, and how characteristics of stone artefact assemblages demonstrate changing patterns of mobility in relation to different ecological conditions.

Simon Holdaway, Matthew Douglass and Patricia Fanning also deal with western New South Wales, addressing some equivalent problems. Recognising that it is important to find the right scale of observation and explanation, they build an argument that variability in hunter–gatherer behaviour in Australia is best seen as a constant and dynamic series of responses to an essentially unpredictable environment. This concept of continual variability provides an alternative to a common view of punctuated points of change alternating with long periods of stasis. There was, rather, a constant negotiation between people and their environment, with diverse adjustments of different types at different times, for which archaeological models derived from outside Australia are not always appropriate.

A fourth example of relating patchy archaeological traces to major environmental changes is provided by David Frankel and Caroline Bird in a study of rockshelters in the Grampians-Gariwerd ranges in south-eastern Australia. Here the juxtaposition of different landforms and shifts over time in their relative productivity and resource reliability set up stresses and attractions that structured demographic movements and social relationships. A diverse array of evidence regarding raw material procurement, tool types, site use, rock art and the environment is integrated to create a model of fluctuations in the place of the ranges in the natural and culturally perceived environment. Frankel and Bird suggest that changes were at times direct responses to climatic circumstances, but at other times activities at sites varied for different reasons, often influenced by accessibility rather than resource availability. Periodic reorganisation in the way the ranges and the surrounding plains were integrated within technological systems and regional land use patterns once again reflects an underlying pattern of flexible responses to natural and social circumstances.

David Thomas, working at a very different scale in a different part of the world and in a much more recent time period, faces similar issues of integrating archaeological data with evidence from other sources. In this case study, which seeks to explain the rapid growth and decline of empires in medieval central Asia, the environment remained constant, but social and technical responses to it changed enormously. Using a multidisciplinary approach in the tradition of the *Annales* School, Thomas demonstrates how the environment played a central role in the imperial trajectory of the Ghūrids, who burst onto the geo-political scene in the middle of the twelfth century. From their origins as a loose agglomeration of chieftains, practising pastoralism and subsistence farming in the mountains of central Afghanistan, the Ghūrids emerged as a major military power. Their initial success and ultimate failure to establish a lasting empire is best understood by considering how both society and environment were transformed by this imperial experiment and affected by historically contingent events. Without external sources of income, this imperial system could not be maintained in this harsh environment, leading to a reversion to a more sustainable, transitory lifestyle.

In the second section, 'Technology and the environment,' chapters deal with different aspects of technology as a key element in driving social responses to the environment. In the first of these Tim Denham reviews a series of historical developments of practices associated with the exploitation of food plants in the Upper Wahgi Valley, Papua New Guinea, over the last 10,000 years. While ethnographic sources document the complex and significant links between people and place in recent times, such specific meanings in earlier prehistory are not accessible to us. The focus in this chapter is therefore on how rather than why people interacted with their world in particular ways, and on the context within which innovations in horticultural practices were adopted. This involves consideration of many intersecting factors, including climatic fluctuations and changes in social and demographic patterns. In addition, other factors lie at the nexus of human–environment relationships, including anthropogenic landscape degradation through forest clearance and soil depletion.

At a finer scale, linkages between agricultural practices (including crop cultivation and animal management) and other technologies (especially pottery manufacture and metallurgy) highlight the intersection between well-established agrarian economies and environments in Steven Falconer and Patricia Fall's comparative study of three sites of the late third and early second millennium BC. Two villages in Jordan display different uses of similar environments: one dedicated to cultivation and animal husbandry, the other also engaged in pottery and metal production, each demonstrating characteristic environmental impacts and archaeological traces. A third, different configuration of technologies is seen in a Cypriot Bronze Age village, where a wooded landscape was managed through a combination of herding, hunting, arboriculture and utilitarian metallurgy. The differences between these pene-contemporary settlements reflect a broader variability in interrelated expressions of household and community behaviour during the development of early complex agrarian societies seen in their specific anthropogenic landscapes.

In another study of prehistoric Cyprus, Jennifer Webb looks at a specific historical development, considering changes in technological and social environments in the second half of the third millennium BC. She contrasts two major archaeologically recognisable cultural entities: an indigenous Late Chalcolithic society dependent on hoe-based agriculture and migrant Early Bronze Age communities with a radically different social and technological system, including the cattle/plough complex. The impact of this relatively sudden introduction of a suite of new technologies gave a significant competitive advantage to the newcomers and presented a major adaptive challenge to the pre-existing population. The two communities can be seen as inhabiting organisationally and ideologically distinct environments— not so much in terms of the actual physical landscape as the perceived or experienced environment. Webb adopts a contextual approach to examine adaptation as a cultural process, identify response mechanisms and model the differential uptake and persistence of technologies and social strategies across the island over the longer term.

Bringing us into the modern world, Peter Davies and Susan Lawrence likewise explore responses to new environments encountered as a result of migration when change occurs within decades rather than over millennia. Colonisation has always been a part of human experience and has brought with it the need to assess and adapt to new environments. The recent past, with its access to documentary and oral evidence, can provide models for understanding some of the processes by which the worldviews and technologies the migrants brought with them were used to ameliorate the effects of unfamiliar environments. Here a combination of archaeological and documentary evidence provides the basis for understanding human–environmental interactions during one of the nineteenth century's great migrations, the gold rush to Australia in the 1850s. Landscape learning models involved in technological responses to water scarcity link several related points of intersection between environment, technology, environmental archaeology and landscape approaches. Archaeological evidence of water management systems is the primary source for examining physical responses to the challenges posed by climate and topography when capturing, storing, transporting and using water for industrial, agricultural and domestic purposes. These allow exploration of issues involved in the adaptation of technology in new environments and industries, changes in water management strategies through time, and the environmental effects of different kinds and scales of water technology. Water is used as the starting point for exploring the complex relationships between people and the natural world, integrating evidence of small-scale, local activities with broader transformations of physical landscapes.

Also drawing on the additional evidence provided by working in the modern world, the two chapters in the third section, under the heading of 'Nature and culture,' present further challenges to some of the simple categories generally used in archaeological analyses by making social systems the point of entry into analyses of technology and environment. Lesley Head and Jennifer Atchison take ethnographic observations of Australian Aboriginal use of yams as the basis for considering the complex relationships between people and plants. Without pre-emptive assumptions or categorisations of human–environment interactions, they ask what kinds of relations are made possible by the material agency of the yam plants themselves, how people might be enrolled into different patterns of care and how particular practices might bundle together. While the food value was certainly important in connecting people and yams, the patterns of care and connection reflect deeper, diverse and complex interactions between plants, people, rocks, soils and digging. From this perspective, any simple separation of nature and culture is replaced by one recognising that they are inextricably entangled, even if inaccessible to prehistorians.

Some related concerns are the subject of Anita Smith's analysis of 'cultural landscapes' since the introduction of this category in the World Heritage system: an important development designed to provide a mechanism for non-Western traditional heritage to be included on the World Heritage List. The large number and diversity of cultural landscapes now inscribed on the list confirm the significance of particular environments or landscapes as a record of human activity, cultural practices and social systems. However, few nominations move beyond the conventional practice of interpreting sites as having either 'natural' or 'cultural' values. Implicit definitions of what constitutes universal heritage value continue to separate the two, and little or no attention is paid to recognising the role of people in modifying and constructing many 'natural' systems, or to how our own Western systems of knowledge construct 'universal' heritage.

The diversity of the chapters in this volume demonstrates the inherent weakness in any attempt to prioritise environment, technology or society. These three factors are all embedded in any human activity. Change in one will result in change in the others, as social and technical changes alter relations with the environment, and indeed the environment itself, and as environmental change drives changes in society and technology. In some circumstances changes will be radical and profound, as in the shift from the Chalcolithic to the Bronze Age in the Mediterranean, or with horticultural innovation in Papua New Guinea. In other circumstances, as in western New South Wales at the end of the Last Glacial Maximum, change will be more subtle as people work to maintain a stable social system in the face of great environmental shifts. As these chapters show, it is possible to consider the relationship between the three factors from different perspectives, but any attempt to consider one or even two in isolation will mean that valuable insights will be missed.

The initial workshop, attended by about half the contributors to this volume, was made possible by a grant from the Faculty of Humanities and Social Sciences, La Trobe University. Janine Major did much to facilitate the event and to assist the project in other ways. Christine Eslick provided valuable editorial assistance, and Wei Ming helped prepare the illustrations for publication.

REFERENCES

Allen, J. & J.F. O'Connell. (ed.) 1995. *Transitions. Pleistocene to Holocene in Australia and Papua New Guinea. Antiquity* 69 (Special number 265).

Anschuetz, K.F., R.H. Wilshusen & C.L. Scheick. 2001. An archaeology of landscapes: perspectives and directions. *Journal of Archaeological Research* 9: 157–211.

Arnold, D.E. 1993. *Ecology and ceramic production in an Andean community.* Cambridge: Cambridge University Press.

Ashmore, W. 2004. Social archaeologies of landscape, in L. Meskell & R. Preucel (ed.) *Companion to social archaeology*: 255–71. Oxford: Blackwell.

Bender, B. 1993. *Landscape: politics and perspectives.* Oxford: Berg.

Butzer, K. 1971. *Archaeology as human ecology.* Cambridge: Cambridge University Press.

Clark, J.G.D. 1957. *Archaeology and society.* 3rd edition. London: Methuen.

Deagan, K. 2008. Environmental and historical archaeology, in E. Reitz, C.M. Scarry & S. Scudder (ed.) *Case studies in environmental archaeology*: 21–42. New York: Springer.

Dincause, D.F. 2000. *Environmental archaeology: principles and practice.* Cambridge: Cambridge University Press.

Dodson, J. (ed.) 1992. *The naïve lands: prehistory and environmental change in Australia and the southwest Pacific.* Melbourne: Longman Cheshire.

Ellen, R. 1982. *Environment, subsistence and system. The ecology of small-scale social formations.* Cambridge: Cambridge University Press.

Evans, J.G. 2003. *Environmental archaeology and the social order.* London: Routledge.

Forde, C.D. 1934. *Habitat, economy and society.* London: Methuen.

Fox, C. 1932. *The personality of Britain: its influence on inhabitant and invader in prehistoric and early historic times.* Cardiff: National Museum of Wales.

Guille-Escuret, G. 1993. Technical innovation and cultural resistance, in P. Lemonnier (ed.) *Technological choices. Transformations in material cultures since the Neolithic*: 214–26. London: Routledge.

Hardesty, D. & D. Fowler. 2001. Archaeology and environmental changes, in C. Crumley, A.E. Deventer & J.J. Fletcher (ed.) *New directions in anthropology and environment: intersections*: 72–89. Walnut Creek: Altamira.

Holtorf, C. & H. Williams. 2006. Landscapes and memories, in D. Hicks & M. Beaudry (ed.) *Cambridge companion to historical archaeology*: 235–54. Cambridge: Cambridge University Press.

Hoskins, W.G. 1965. *The making of the English landscape.* London: Hodder and Stoughton.

Kirch, P. 2005. Archaeology and global change: the Holocene record. *Annual Review of Environment and Resources* 30: 409–40.

Knapp, A.B. & W. Ashmore. 1999. Archaeological landscapes: constructed, conceptualized, ideational, in W. Ashmore & A.B. Knapp (ed.) *Archaeologies of landscape: contemporary perspectives*: 1–30. Oxford: Blackwell.

Lemonnier, P. 1993. Introduction, in P. Lemonnier (ed.) *Technological choices: transformations in material cultures since the Neolithic*: 1–35. London: Routledge.

Leone, M. 1984. Interpreting ideology in historical archaeology: using the rules of perspective in the William Paca Garden in Annapolis, Maryland, in D. Miller & C. Tilley (ed.) *Ideology, power and prehistory*: 25–35. Cambridge: Cambridge University Press.

Miller, H.M.-L. 2007. *Archaeological approaches to technology.* Burlington: Academic Press.

Mrozowski, S. 2006. Environments of history: biological dimensions of historical archaeology, in M. Hal & S. Silliman (ed.) *Historical archaeology*: 23–41. Oxford: Blackwell.

Mulvaney, D.J. & J. Golson. (ed.) 1971. *Aboriginal man and environment in Australia.* Canberra: Australian National University Press.

Salisbury, R.R. 1962. *From stone to steel: economic consequences of technological change in New Guinea.* Parkville: Melbourne University Press.

Sauer, C. 1925. The morphology of landscapes. *University of California Publications in Geography* 2: 19–54.

Thomas, W.L. & C.O Sauer. (ed.) 1956. *Man's role in changing the face of the earth.* Chicago: University of Chicago Press.

Part I

Responses to Environments

2 Perspectives on Global Comparative Hunter–Gatherer Archaeology
Glacial Southwest Tasmania and Southwest France

Richard Cosgrove, Jean-Michel Geneste, Jean-Pierre Chadelle and Jean-Christophe Castel

INTRODUCTION

Studies of Late Pleistocene hunter–gatherers on different continents have attempted to compare their archaeological records in the hope of revealing global temporal and spatial variability (Soffer 1987; Gamble & Soffer 1990; Veth et al. 2005; Smith & Hesse 2005). However, incomplete or uneven chronological and palaeoecological frameworks have limited the ability of most studies to compare archaeologies in commensurate ways. In one major attempt to do this, Gamble and Soffer (1990: 15) used the Last Glacial Maximum as a point of comparison but identified serious limitations in the data because of the different research agendas and the use of conflicting spatial and chronological scales. They argued that any observed variability could be the product of incompatible data sets rather than significant differences in human behaviours (1990: 15) and suggested that if global comparisons are to be successful in illuminating aspects of human behavioural variability, it is crucial that standardisation of both temporal and spatial scales be in place for cross-cultural studies (1990: 8). This approach has been successful in the natural sciences, particularly in studies of past global climates (e.g. Gasse et al. 1997; Shulmeister et al. 2006).

In this study the comparison is between Late Pleistocene southwest France and southwest Tasmania. These areas were chosen because they are where people were under increased selective pressure, where social and ecological tensions were at their most intense during the prevailing glacial climatic conditions (Wobst 1990:339). Significantly, the two regions are rich in archaeological remains and appear to reflect the ebb and flow of human populations during glacial and interglacial cycles (Cosgrove 1995a, 1995b; Holdaway & Porch 1995; Gamble et al. 2004; Gamble et al. 2005).

The presence of these records implies variable distributions and densities of human populations and resources across landscapes. These, in turn, encourage questions about how climate affects cultural trajectories and recolonisation processes at a global scale. Although both groups of people were clearly behaviourally modern (Cosgrove et al. in press), they shared little in the way of a common cultural or even genetic history (Rasmussen et al. 2011) but, for different reasons, abandoned aspects of long-term economic strategies focused on cave sites.

Identifying the appropriate archaeological frameworks within which to examine the archaeological variability between the two regions is a crucial step. At present there is still a dearth of appropriate frameworks to apply to questions of regional variability, particularly on a comparative global basis. Most global comparative archaeologies have used diachronic approaches to examine differences and similarities in complex societies (Peregrine 2004). The common analytical units of comparison are usually chiefdoms, city-states or the origins of food production. In European Palaeolithic studies, the common unit of comparative analysis has been stone tool types organised into analytical units thought to reflect cultural groupings (Gamble 1993: 164; Shea 2011).

Archaeologists have long speculated about the common problems faced by humans under rigorous ecological circumstances and whether analogous behaviours in different parts of the world, occupied at similar times, can be traced in the archaeological record (Gamble & Soffer 1990). However, eliminating methodological variability in these records is a challenge, especially in assemblages that have their own inherent biases and approaches structured by historical contingency. A useful method applied to the Australian archaeological record is the quantification of discard rates per unit time or some other corrected measure, the assumption being that these fluctuations reflect differential human occupation intensities. This method can also standardise or dampen the effects of disparate sedimentation rates that so often affect the structural properties of archaeological records. The aim of this chapter is to analyse first-order faunal and technological data that might identify cognate discard patterns in the two separate regions, based on chronologically commensurate data sets.

BACKGROUND

The chapter examines earlier comparisons between the rich Upper Palaeolithic cave sites in southwest France and the Tasmanian Late Pleistocene sites (Kiernan et al. 1983; Jones 1984, 1990: 281, 290; Cosgrove et al. 1990; Cosgrove & Allen 2001; Cosgrove & Pike-Tay 2004). One observation by Jones (1984: 59) was that "if there are common elements in these histories, they may tell us something fundamental about human behaviour." These initial suggestions emphasised many superficial similarities, such as the richness in the archaeological cave deposits. It was equally apparent that there are many differences that reflect varied human behavioural responses in the two regions to broadly similar climatic conditions over the same time period.

There are general similarities and differences that can readily be recognised. For example, the ice age settlement of southwest Tasmania and southwest France (40,000–10,000 years ago) was part of the global spread of anatomically modern humans out of Africa. Their sites are filled with enormous quantities of stone tools and bone from a limited array of animal prey species: Bennett's wallaby and wombat in Tasmania (Cosgrove 1999), reindeer and horse in France. However, each possessed a very different set of material culture associated with hunting and gathering activities within different ecological settings (Cosgrove & Allen 2001: 399) and these should reflect independent adaptations to commensurate environmental fluctuations.

It is these characteristics that raise questions about how people at this time adjusted their social and economic behaviours to these challenging climatic perturbations. Here we focus on an initial analysis of the French site of Combe Saunière as a starting point for comparison with the Tasmanian evidence.

COMBE SAUNIÈRE

Combe Saunière, in the Isle River valley in the Dordogne region of southwest France, is the focus of this analysis (Figure 2.1). This limestone rockshelter has been the target of excavation and analysis for over 20 years. It is located in a low cliff in the Isle River valley about 30km northeast of Périgueux and measures *c.* 20m long x 9m deep, with cultural layers about 5m deep. Geneste and Chadelle excavated the site between 1978 and 1996, identifying a sequence of human occupations spanning *c.* 90,000 to 14,000 BP (Geneste & Plisson 1986). The stratigraphy is well understood in relation to the archaeology and over 70 radiocarbon and electron spin resonance (ESR) dates have been obtained from five layers. The basal levels are composed of limestone and, in some places, kastic sands that are overlain by a sequence of cultural layers from Mousterian, Chatelperronian, Aurignacian, Gravettian and Solutrean, with minor components of Magdalenian at the top (Figure 2.2). A major faunal and stone technological study of the Solutrean layers has been undertaken by Geneste and Plisson (1986), Geneste and Maury (1997) and Castel et al. (2006), but the other layers have not been systematically analysed. Importantly, the excavation techniques are similar to those used at the Tasmanian sites, with 50cm cells dug in spits of *c.* 2.5cm.

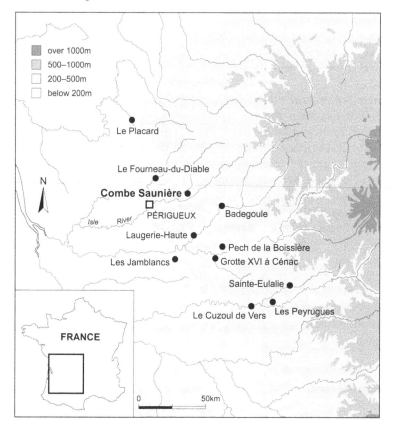

Figure 2.1   Location of Combe Saunière, southwest France. Other sites have well-preserved fauna dating to the Solutrean

Previous analysis (Castel et al. 2006; Geneste et al. 2008) focused on the spatial distribution of faunal remains from the Solutrean layers only. They noted the most numerous animals in these layers were reindeer (*Rangifer tarandus*; number of identified specimens present (NISP) = 4998; 70%), horse (*Equus caballus*; NISP = 513; 7%), fox (*Vulpes vulpes*; NISP = 560; 7.8%) and hare (*Lepus* sp.; NISP = 303; 4.2%). The fox and hare were considered by Castel et al. (2006) to be predator and its prey respectively, while the reindeer and horse were considered human prey. It was suggested that foxes became embedded in the deposits because they died in their burrows inside the shelter. Lower NISP were recorded for other human prey such as bovids (NISP = 92; 1.2%), chamois (NISP = 83; 1.1%), saiga antelope (NISP = 44; 0.6%) and red deer (NISP = 43; 0.6%).

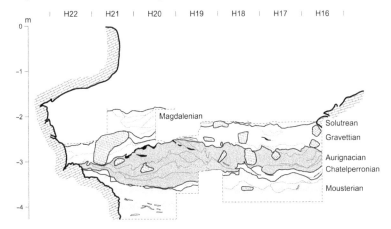

Figure 2.2   The stratigraphy of Combe Saunière. Bronze Age and historic layers omitted for clarity

Most of the stone projectile points (97%) found in the Solutrean layers were broken. Of the end points 20% were probably introduced into the deposits embedded in prey carcasses, while 49% were broken during maintenance and repairs to spear armatures (Geneste & Plisson 1993: 130–1). Hunting of reindeer occurred mostly in late winter, early spring and into summer (Castel et al. 2006: 143; Geneste et al. 2008), and systematic butchery was noted, based on the presence of specific reindeer body parts (Castel et al. 2006: 143). Bone and antler from the reindeer were used for such secondary purposes as awls, retouching pieces, pins and eyed needles, projectile points, chisels, a spear thrower hook and a drilled baton (Castel et al. 2006: 149). Some items had been transported long distances, including mammoth ivory, perforated teeth and marine shell from the Atlantic and Mediterranean coasts. The presence of perforated ibex incisors suggests their introduction as ornaments, as the natural range of the ibex would have been to the southeast, in the mountains of the Massif Central. Ninety per cent of the raw materials used for lithic production were exploited within a 5km radius of the site. They were mainly retouched tools, although unworked material in the form of blanks was imported from the north, south and southeast, suggesting social connections beyond the local geographic zone (Geneste et al. 2008).

The Solutrean occupation of Combe Saunière has been argued to reflect use by mobile hunter–gatherers moving between sites and resource areas within the region. It probably functioned as a satellite site to larger settlements that contained more diverse material culture and proportionally fewer projectile points (Geneste et al. 2008: 233). Unlike some bigger sites, for example Laugerie Haute, in the Vézère River valley, Combe Saunière did not contain any portable or parietal art signifying different functions or ranking within the cultural system at the time (Geneste et al. 2008: 232). The Solutrean phase of Combe Saunière is interpreted as a temporary hunting location, where butchery of prey animals and the maintenance of hunting equipment such as shouldered points took place.

How this may have changed over *la longue durée* is unknown, but preliminary temporal patterns identified in this study suggest rising and falling tempos of occupation over the millennia. It has been suggested that human population shifts detected during the ice age of western Europe indicate fluctuating north–south environmental clines closely linked to temperature changes during the glacial period (Drucker et al. 2003; Gamble et al. 2004; Stevens et al. 2008). Gamble et al. (2004) used existing numbers of calibrated radiocarbon dates between 30,000 cal BP and 6000 cal BP as a proxy for population size and movement. Changes in the frequency of ^{14}C dates suggested a range of responses by humans to variable temperatures. Of particular interest here is the greater representation of Gravettian and Solutrean radiocarbon pulses during the Last Glacial Maximum. Gamble et al. (2004) suggest that these represent people moving north from the Iberian Peninsula into glacial refugia during this cold period. They argue that the increase in ^{14}C dates reflects small human populations expanding into areas north of the Pyrenees into the Aquitaine Basin, where the site of Combe Saunière was occupied. Similarly, in Tasmania fluctuating numbers of ^{14}C dates and artefact discard rates have been argued to represent variable pulses of human occupation within Late Pleistocene sites (Cosgrove 1995b; Holdaway & Porch 1995; Smith & Sharp 1993: 37–59).

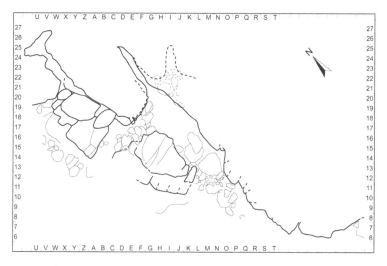

Figure 2.3   Plan of Combe Saunière showing squares H16–H21

Thus, one of the questions we address here is whether overall intensities of human occupation that correlate with the radiocarbon signal detected (Gamble et al. 2004) can be identified at shelters such as Combe Saunière using the <>2cm lithic and faunal debris in each cultural phase. A preliminary analysis is undertaken here, using samples from Trench H in the central area of Combe Saunière (Figure 2.3).

Methods

Trench H was chosen because it best represented the stratigraphic units and had an uninterrupted sequence from the rear wall to the front of the shelter. All the <2cm material in maximum dimension from the sub-squares H16A/C to H21A/C was analysed. Approximately 50 spits (*décapage*) that date between about 40,000 and 13,000 BP, from nine 50cm squares in H16A/C through to H21A/C, were studied (Table 2.1). A total of 52,368 bones and 29,425 lithics <2cm in size were counted and weighed in this first-order study. The analyses of the Mousterian layers have not been finished, so this period was not included in this study. These frequencies were then combined with all stone tools and faunal remains >2cm plotted during the excavations in 3D in each spit within each cultural phase, totalling 87,308 objects. A more thorough faunal analysis of *c.* 11,000 bones from two 50cm cells, H18A and H17A, was completed in 2009. The results provide an initial indication of species, body-part selection, butchery strategies and prey selection in the cultural layers.

The cultural phases used as bases for the comparative analytical units in this study generally matched the sequences in southwest Tasmania. These were from the Chatelperronian to Magdalenian. The chronological range for each phase was estimated using 54 radiocarbon and ESR dates sampled from between *c.* 1.6m and 4.4m below the datum (Table 2.1).

Table 2.1   The age range for each cultural phase and corresponding time span is shown, as determined from the dating at Combe Saunière. Calibrated ages are calculated from CalPal online program (Danzeglocke et al. 2010)

Cultural phase (analytical units)	Age (^{14}C and ESR)	Approximate time span (years BP)	Calibrated age cal BP (CalPal)
Mousterian	84,400–68,900	15,000	
Chatelperronian	42,900–33,000	9,900	
Aurignacian	35,500–27,700	7,800	37,700–32,200
Gravettian	27,800–21,600	6,200	32,400–27,400
Solutrean	22,200–17,400	4,800	26,400–20,900
Magdalenian	17,700–13,900	3,800	21,100–17,100

Preliminary Results

Lithics

The relative frequencies and discard per 1000 years of all plotted artefacts and faunal remains from all the squares within each cultural phase were calculated and are shown in Figure 2.4a. The overall trends show an increasing number of lithics and fauna discarded through time until the Solutrean phase, after which values decline into the Magdalenian phase. This trend is due to the small representative sample surviving in the upper layers (Castel et al. 2006). As the Magdalenian period is also not well represented at Combe Saunière, the decline may reflect either lower levels of occupation or settlement elsewhere in the more protected river valleys to the west and south until the amelioration of climate, *c.* 13,000 BP (Langlais et al. 2012). The site also has evidence of historic occupation, which may have led to the removal of some of the upper layers.

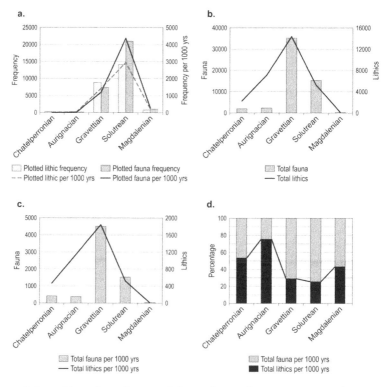

Figure 2.4   a. Plotted artefacts and fauna from all squares at Combe Saunière; b. Total frequency of plotted and sieved fauna and lithics discarded in sub-squares H16A/C–H21A/C from each cultural phase; c. Total frequency of plotted and sieved fauna and lithics discarded in sub-squares H16A/C–H21A/C from each cultural phase per 1000 years; d. Percentage of total lithic and faunal remains discarded per 1000 years by cultural phase in Trench H, sub-squares H16A/C–H21A/C

Figures 2.4b and 2.4c illustrate the patterns found when the <2cm material is combined with the >2cm plotted artefact and faunal remains from sub-squares H16A/C–H21A/C only. There are apparently relatively low levels of early occupation in this area of the cave compared to that in the later Gravettian and Solutrean periods, although some caution is needed when reading the graphs, as not all material below the Gravettian layers was plotted. Nevertheless the distributions are variable, with stone debris present in increasingly high numbers in the Aurignacian (n = 7080) periods. It is in the Gravettian (n = 14,362) that the largest increases occur, with an apparent decrease during the Solutrean (n = 5218). This appears contrary to the overall pattern of plotted items in Figure 2.4a and may reflect increased stone reduction in the Gravettian phase, producing higher amounts of smaller debitage per unit time.

The percentage of lithic material discarded from each time period declines but with a corresponding increase in faunal remains (Figure 2.4d). This pattern is calculated on the basis of discard per 1000 years from the Chatelperronian (lithics = 53%; fauna = 47%) and Aurignacian (lithics = 75%; fauna = 25%) to the Gravettian (lithics = 27%; fauna = 73%) and Solutrean cultural phases (lithics = 22%; fauna = 78%). It suggests an increasing interest in faunal exploitation towards the coldest part of the Last Glacial Maximum and correlates with a focus on the development of specialised hunting technology, such as Solutrean shouldered points.

There is also a corresponding increase in a range of typologically distinct stone tools, as reflected in those plotted in 3D in each spit in squares H16A/C–H21A/C. Although there is a range of different typological categories, these have been amalgamated for ease of analysis. As might be expected, the distribution of each tool class is distinct between the cultural layers. The most notable change in tool types is reflected in scrapers (Aurignacian = 14%; Gravettian = 79%; Solutrean = 7%), retouched pieces (Aurignacian = 4%; Gravettian = 70%; Solutrean = 26%), burins (Aurignacian = 2%; Gravettian = 91%; Solutrean = 7%), Solutrean shouldered points (Aurignacian = 0%; Gravettian = 10%; Solutrean = 90%) and Gravette points (Aurignacian = 0%; Gravettian = 80%; Solutrean = 10%).

Other notable differences are seen in the presence of notched fragments that make up 89% in the Gravettian period but only 11% in the Solutrean. Blades occur in all cultural periods but with higher percentages in the Gravettian (66%) and lower in the Aurignacian (4%) and Solutrean (30%). Other distinctive tools occur only in their respective 'cultural' groups. For example, Chatelperron points (100%; n = 2), Perigordian *flechettes* (100%; n = 5), Dufor bladelets (88%; n = 7) and Solutrean points (90%; n = 35) are found in the Aurignacian, Gravettian and Solutrean analytical units. The concentration of tool types in the Gravettian appears to accompany the increases in <2cm debitage and supports the notion of increased knapping and stone reduction. The Solutrean seems to be a period preoccupied with specialist tools, such as the relatively high number of shouldered points, the sine qua non of this cultural phase.

When all plotted stone artefact material is combined, there are apparent decreases in average weights over time, from the Aurignacian to the Solutrean. An ANOVA one-way analysis of variance on 463 tools (F ratio = 34.5, Probability > F = <0.001) shows a significant difference of mean weight between the Aurignacian (mean = 36g, SD = 37.3), the Gravettian (mean = 7.8g, SD = 14.4) and the Solutrean (mean = 3.8g, SD = 11.7). There is also some evidence from the H16A/C–H21A/C squares to suggest that there are also significant differences in the dimensions of tools common to the cultural divisions. For example, the lengths of retouched pieces become smaller between the Aurignacian (mean = 70mm), the Gravettian (mean = 36mm) and the Solutrean (mean = 14mm), although the above results need further investigation through an analysis of a larger sample within different squares from across the site.

Fauna

Preliminary studies of the human prey show the overwhelming dominance of medium mammals in every cultural period from sub-squares H18A and H17A. The key human prey species are reindeer (NISP = 379; 32%) and medium mammal (NISP = 585; 49%) from a total of 1197 NISP. The latter category is almost certainly made up of fragmented reindeer bones that bring the proportion of these animals to at least 81% as the primary human prey. Reindeer is also the major prey animal in all the cultural phases, with between 78.6% and 87.7% of the total fauna. The medium ungulate (NISP = 91; 8%), bovid (NISP = 15; 1.2%), chamois (NISP = 15; 1.2%), deer (NISP = 8; 0.6%), horse (NISP = 47; 4%) and large mammal/ungulate (NISP = 53; 4.7%) appear to be only minor components of the hunted prey. The dominance of reindeer parallels results from other studies in the region that suggest reindeer hunting may have played a major economic role in human adaptation during glacial conditions (Boyle 1990; Castel et al. 2005; Castel 2010; Castel et al. 2006; Grayson & Delpech 2004; Mellars 2004; Stevens et al. 2008). They provided meat, skins, fuel, and bone and antler tools, as well as decorative ornaments (Castel et al. 2006).

Overall the faunal distribution is similar to the lithic patterns, with the frequency of combined <>2cm lithic material peaking in the Gravettian period. However, these differences cannot be accounted for by more fragmented prey animal bones in the Gravettian, since there is no significant difference in the lengths of bones between any of the cultural phases as measured by NISP and NISP/1000 years. Thus increases in the Gravettian cannot be accounted for by increased fragmentation based on total length of reindeer bones, as there is no significant difference between the cultural periods (F ratio = 1.26; Probability > F = 0.28). It may indicate greater overall numbers of reindeer being brought onto the site at these times or different bone treatment compared to that of the other phases. To confirm this as a general trend, a larger sample size from across the site is needed.

There is also consistent reindeer body-part exploitation, with most time periods containing a very high proportion of long bone elements. In the Aurignacian it is 75%, Gravettian 74% and Solutrean 67% out of a total assemblage of 805 NISP. Of interest are the very low numbers and percentages of the axial elements of between 8% and 5% in all cultural phases, with even lower frequencies and percentages of pectoral and pelvic girdles of between 0.2% and 2.2%. The results are similar to those of Castel et al. (2006), who found these body parts were under-represented in the Solutrean layers at Combe Saunière.

Indeed, there is repeated representation of the same bone elements in all the cultural phases examined in sub-squares H18A and H17A. The most numerous are the femora, humeri, mandible, metatarsals, ulnae, radii/ulnae and tibiae (NISP = 815), but ribs and vertebrae are all under-represented in the assemblage, supporting notions that these bones were burnt or not brought back to the site (Castel et al. 2006). The over-representation of specific meat-bearing bones in many archaeological sites has been interpreted as selective body-part butchery, in this case reindeer hind and forequarters. The targeted exploitation of marrow and the manufacture of tools from the reindeer long bones and other body parts such as antler have been identified in detailed studies of the Solutrean layer, a pattern that now appears common in the other cultural periods (Castel et al. 2005). Seasonal exploitation of reindeer has been identified as a spring/summer event at Combe Saunière during the Solutrean period (Castel et al. 2006). Given that ungulates such as reindeer lose body condition during the winter and early spring, it may be suggested that these long bones were deposited mainly during the summer months from kills in the nearby valleys (Geneste et al. 2008; Garvey 2011).

SCALES OF COMPARISON

We now turn our attention to a preliminary comparison of the patterns identified at Combe Saunière with those of the southwest Tasmanian sites. There are at least three points of comparison that can be made between the two regions: palaeoenvironment, technology and faunal exploitation.

Palaeoenvironments

Despite lying between the 42nd and 43rd degrees latitude south and north, one of the obvious differences between southwest Tasmania and southwest France is their ecological structures. It is clear that between 40,000 and 13,000 BP, southwest France experienced fluctuating climates (Stevens et al. 2008) that lowered sea levels, increasing the area of western Europe considerably (Lambeck & Bard 2000; Lambeck 2004). Grassland would have been extensive, and nitrogen isotope studies on reindeer reveal the presence of continuous permafrost conditions as far south as the current northern French–German border during the height of the Last Glacial Maximum (Stevens et al. 2008). Discontinuous and sporadic permafrost conditions probably existed in the region around Combe Saunière and southwest France generally between *c.* 27,000 and 15,000 BP. It would suggest the existence of tundra-steppe conditions throughout the human occupation of Combe Saunière. For much of that time it would have been extremely arid (Drucker et al. 2003; Stevens et al. 2008), until after 13,000 BP, when temperatures started to rise and conditions ameliorated (Stevens et al. 2008; Langlais et al. 2012).

Tasmania experienced similar global climate impacts, but because of its more maritime position as a peninsula of Sahul (New Guinea–Australia–Tasmania) the climate was probably relatively mild, with more oceanic influences. The glaciers were small and restricted to the highest peaks (Mackintosh et al. 2006), and pollen records suggest that grass and savannah were extensive in eastern and north-western Tasmania, while herbfields and grassland patches occupied the lower valleys where human occupation appears to have been concentrated (Cosgrove & Allen 2001; Colhoun & Schimeld 2012). Unlike southwest France, temperature depressions in Tasmania were not as severe. Estimates of mean annual temperatures in Europe were less than—8°C (Stevens et al. 2008) and variations in annual temperature were large, probably reaching between—25°C and—20°C in the winter months. In Tasmania during the height of the glacial annual average temperatures were about 4°C, based on geomorphic, pollen and emu eggshell AAR studies (Cosgrove 1995a: 76; Colhoun & Schimeld 2012). No evidence for permafrost or extreme aridity during the last glacial was found in western Tasmania (Fitzsimons & Colhoun 1989: 352), although it was dry and cold.

Technology

A distinguishing feature of the European Upper Palaeolithic is the abrupt technological changes that took place and are used to define cultural phases. As described above, there is clear evidence at Combe Saunière for distinct categories of stone projectile point, scrapers and burins that characterise each cultural phase. These have been interpreted as providing rapid solutions to the rather high-frequency, high-amplitude temperature fluctuations experienced by hunters in this region between 40,000 and 13,000 BP (Taylor et al. 1993; Langlais et al. 2012).

Although there are some distinctive stone tool types in the Tasmanian Late Pleistocene, such as small denticulate flakes and 'thumbnail' and end scrapers, only small and gradual adjustments to technology occurred over 25,000 years of human occupation (Cosgrove 1999; Holdaway 2004).

No stone projectile points or blades have been identified in any of the assemblages, although organic material such as wooden spears, baskets and waddies were probably part of a flexible organic technology. In hunting weapons, bone point armatures (Webb & Allen 1992) were the functional equivalent of the bone-, stone- and ivory-tipped spears of southwest France. Despite the possibility of hafting in Tasmania, no evidence of material such as resin has been found microscopically adhering to these tools. There is also some evidence that the bone points were used as awls to pierce wallaby skins to make clothing (Webb & Allen 1992).

Raw material distribution in southwest Tasmanian sites mirrors the use of local stone for artefacts at Combe Saunière. In Tasmania, most raw materials occur less than 1–2km from the sites, although some small quantities of Darwin glass and blue/grey chert are sourced from over 100km away from occupied cave sites (Cosgrove 1999). At Combe Saunière some stone artefact raw materials used specifically for well-designed hunting equipment had travelled up to 200km from northern sources (Geneste et al. 2008: 231). Significantly, discard rates of <>2cm lithic material at Combe Saunière have similar trends to those of the lowland Tasmanian cave sites of Kutikina and Warreen, with increases between about 27,000 and 20,000 BP. In the upland cave sites of Bone and Nunamira most discard appears to happen between *c*. 29,000–24,000 BP and *c*. 17,000–13,000 BP. The parallel patterns from Combe Saunière suggest that temperature fluctuations during the Last Glacial Maximum had a considerable effect on occupation intensities at different times and altitudes, where some sites were abandoned or had relatively reduced occupation levels.

Faunal Exploitation

The distribution of bone discards parallels the patterns in lithic material. At Combe Saunière, where reindeer is the major human prey, the densest accumulation of bones appears to be between 28,000 and 17,000 BP. Likewise, in southwest Tasmania the lowland cave sites of Warreen and Kutikina have similar temporal patterns, with most of the prey remains discarded between 27,000 and 15,000 BP. This suggests that independent human responses to the severe glacial conditions prevailing at this time in both hemispheres were planned to lower economic risk and uncertainty (Halstead & O'Shea 1989). At Combe Saunière the reindeer body parts are dominated by long bones, particularly the hindquarter and forequarter. In Tasmania the Bennett's wallaby was the principal prey animal, making up 75–90% of all bones analysed. The lower limb bones were heavily processed to extract the marrow, a valuable source of fats.

It is clear that both these animals provided a secure source of nutrients during the coldest part of the glacial and were especially important because of the low amount of plant carbohydrates available in this period (Cosgrove et al. 1990; Garvey 2011). As noted above, at Combe Saunière secondary products from the reindeer were produced from skins, bones and antler (Castel et al. 2006), while in southwest Tasmania bone points and spatulas made on wallaby fibulae were a component of the bone technology. These were probably used to make cloaks or clothing from wallaby skins (Cosgrove 1999).

Seasonal exploitation of prey animals has been identified in both regions. Combe Saunière, Bone and Nunamira caves have similar patterns, with hunting occurring between late winter/early spring and summer/late autumn (Castel et al. 2006; Pike-Tay et al. 2008). In the lowland cave sites of Kutikina and Warreen, most hunting was done in winter. The reasons for these different patterns appear to be clear. Reindeer suffer fat depletion towards the end of winter and after the rut, so summer seems to be the time for optimal returns (Garvey 2011). They are hunted on an intercept basis along strategic migration routes. Wallabies, however, are relatively sedentary (Cosgrove et al. 1990), with small home ranges. It is argued that hunters killed wallaby after travelling to small patches of grassland to which these animals are ecologically tethered (Cosgrove & Allen 2001). In contrast to reindeer, wallabies do not suffer seasonal fat depletion, and so it appears that human movement into the Tasmanian uplands in summer was not exclusively to hunt wallabies in optimal condition, as they would have been available all year round. Other activities were probably embedded in these movements, although the need to procure lithic materials was probably not one of them.

What is clear is that at the end of the glacial period, about 13,000 BP, both economic strategies came to an abrupt end with differing responses to social and ecological conditions after the Last Glacial Maximum. In Europe we see people becoming more settled through Mesolithic times, represented by the Azillian period in southwest France at places such as Mas d'Azil in the Ariège, northern Pyrenees (Meiklejohn et al. 2010). In Tasmania, as the earth warmed after 13,000 years ago, trees invaded the southwest, covering the grassland patches and driving out the wallabies that had for so long formed the economic basis of human settlement in the region. With the disappearance of the wallabies, the people moved to other locations in the north and east.

The time also heralded the rise of sea levels, with the drowning of the continental shelf in western Europe and, more significantly, the cutting of the Bass land bridge that for over 25,000 years had connected Tasmania as part of Sahul. The Tasmanian Aboriginal culture persisted through the next 10,000 years as one of the longest continuous hunter–gatherer traditions anywhere in the world.

ACKNOWLEDGMENTS

This project was funded by the Australian Academy of Science, Scientific Visits to Europe Program; Ian Potter Foundation Travel Grant; and University of Bordeaux. We wish to thank Jean-Jacques Cleyet-Merle, Director, National Museum of Prehistory, Les Eyzies-de-Tayac for supporting the project. We also thank Dr Stephane Madelaine and André Morala for assistance with bone identification and organising access to the faunal collections. One of us (RC) was a recipient of a University of Bordeaux Archaeology Fellowship, and he would like to thank Dr Jean-Philippe Rigaud (then Director, Institut du Quaternaire), Professor Paola Villa (then Head of Department), Dr Françoise Delpech and Dr Michel Lenoir for assistance while at Centre Bordes. We thank Professor David Frankel, Dr Jenny Webb and Dr Susan Lawrence for the invitation to present our results to the symposium, *Intersections and Transformations: Archaeological Studies in Environment and Technology*, La Trobe University, November 2010.

REFERENCES

Boyle, K.V. 1990. *Upper Palaeolithic faunas from south-west France* (British Archaeological Reports International Series 557). Oxford: Tempus Reparatum.

Castel, J-C. 2010. Comportements de subsistence au Solutréen au Badegoulien d'après les faunes de Combe Saunière (Dordogne) et du Cuzoul de Vers (Lot). Unpublished PhD dissertation, University of Bordeaux.

Castel, J-C., J.-P. Chadelle & J.-M Geneste. 2005. Nouvelle approche des territoires solutréens du Sud-Ouest de la France, in J. Jaubert & M. Barbaza (ed.), *Territoires, déplacements, mobilité, échanges durant la Préhistoire*, 126e congrès du CTHS, Toulouse, 2001: 279–94. Paris, Editions du CTHS.

Castel, J-C., D. Liolios, V. Laroulandie, F-X. Chauvière, J-P. Chadelle, A. Pike-Tay & J-M. Geneste. 2006. Solutrean animal resource exploitation at Combe Saunière (Dordogne, France), in M. Maltby (ed.) *Integrating zooarchaeology*: 138–52. Oxford: Oxbow.

Colhoun, E.A. & P.W. Schimeld. 2012. Late-Quaternary vegetation history of Tasmania from pollen records, in S. Haberle & B. David (ed.) *Peopled landscapes: archaeological and biographic approaches to landscapes*: 297–328 (Terra Australis 34). Canberra: ANU E Press.

Cosgrove, R. 1995a. *The illusion of riches: scale, resolution and explanation in Tasmanian Pleistocene human behaviour* (British Archaeological Reports International Series 608). Oxford: Tempus Reparatum.

Cosgrove, R. 1995b. Late Pleistocene behavioural variation and time trends. *Archaeology in Oceania* 30: 83–104.

Cosgrove, R. 1999. Forty-two degrees south: the archaeology of Late Pleistocene Tasmania. *Journal of World Prehistory* 13: 357–402.

Cosgrove, R. & J. Allen. 2001. Prey choice and hunting strategies in the Late Pleistocene: evidence from southwest Tasmania, in A. Anderson, S. O'Connor & I. Lilley (ed.) *Histories of old ages: essays in honour of Rhys Jones*: 397–429. Canberra: Coombs Academic Publishing, Australian National University.

Cosgrove, R. & A. Pike-Tay. 2004. The Middle Palaeolithic and late Pleistocene Tasmania hunting behaviour: a reconsideration of the attributes of modern human behaviour. *International Journal of Osteoarchaeology* 14: 321–32.

Cosgrove, R., J. Allen & B. Marshall. 1990. Palaeoecology and Pleistocene human occupation in south central Tasmania. *Antiquity* 64: 59–78.

Cosgrove, R., A. Pike-Tay & W. Roebroeks. In press. Tasmanian archaeology and reflections on modern human behaviour, in R. Dennell & M. Porr (ed.) *East* of Africa. *South Asia, Australia and human origins.* Cambridge: Cambridge University Press.

Danzeglocke, U., O. Jöris & B. Weninger. 2010. CalPal-2007online. Available at: http://www.calpalonline.de/ (accessed 1 May 2012).

Drucker, D.G., H. Bocherens & D. Billiou. 2003. Evidence for shifting environmental conditions in southwestern France from 33,000 to 15,000 years ago derived from carbon-13 and nitrogen-15 natural abundances in collagen of large herbivores. *Earth and Planetary Science Letters* 216: 163–73.

Fitzsimons, S.J. & E.A. Colhoun. 1989. Pleistocene clastic dykes in the King Valley, western Tasmania. *Australian Journal of Earth Sciences* 36: 351–63.

Gamble, C. 1993. *Timewalkers: the prehistory of global colonization.* Phoenix Mill: Allan Sutton.

Gamble, C. & O. Soffer. 1990. Introduction. Pleistocene polyphony: the diversity of human adaptations at the Last Glacial Maximum, in C. Gamble & O. Soffer (ed.) *The world at 18,000 BP. Volume 2: low latitudes*: 1–23. London: Unwin Hyman.

Gamble, C., W. Davies, P. Pettitt & M. Richards. 2004. Climate change and evolving human diversity in Europe during the last glacial, in K.J. Willis, K.D. Bennett & D. Walker (ed.) The evolutionary legacy of the ice ages. *Philosophical Transactions of the Royal Society Biological Sciences* 359: 243–54.

Gamble, C., W. Davies, P. Pettitt, L. Hazelwood & M. Richards. 2005. The archaeological and genetic foundations of the European population during the Late Glacial: implications for 'agricultural thinking.' *Cambridge Archaeological Journal* 15(2): 193–223.

Garvey, G. 2011. Bennett's wallaby (*Macropus rufogriseus*) marrow quality vs. quantity: evaluating human decision-making and seasonal occupation in late Pleistocene Tasmania. *Journal of Archaeological Science* 38: 763–83.

Gasse, F., F. Kroepelin & S. Oldfield (ed.). 1997. *PEP III: The Pole-Equator-Pole transect through Europe and Africa*: 1–64 (PAGES workshop report series 97:2). Bern, Switzerland: PAGES International Project Office.

Geneste, J-M. & S. Maury. 1997. Contributions of multidisciplinary experimentation to the study of upper Paleolithic projectile points, in H. Knecht (ed.) *Projectile technology*: 165–189. New York: Plenum Press.

Geneste, J-M., J-C. Castel & J-P. Chadelle. 2008. From physical to social landscapes: multidimensional approaches to the archaeology of social place in the European Upper Palaeolithic, in B. David & J. Thomas (ed.) *Handbook of landscape archaeology*: 228–36. Walnut Creek: Left Coast Press.

Geneste, J-M. & H. Plisson. 1986. Le Solutréen de la Grotte de Combe Saunière 1 (Dordogne) première approche palethnologique. *Gallia Préhistoire* 29(1): 9–27.

Geneste, J-M. & H. Plisson. 1993. Hunting technology and human behaviour: lithic analysis of Solutrean shouldered points, in H. Knecht, A. Pike-Tay & R. White (ed.) *Before Lascaux: the complex record of the early upper Paleolithic*: 117–35. Boca Raton: CRC Press.

Grayson, D. & F. Delpech. 2004. Pleistocene reindeer and global warming. *Conservation Biology* 19(2): 557–62.

Halstead, P. & J. O'Shea. 1989. Introduction: cultural responses to risk and uncertainty, in P. Halstead & J. O'Shea (ed.) *Bad year economics: cultural responses to risk and uncertainty*: 1–7. Cambridge: Cambridge University Press.

Holdaway, S. 2004. *Continuity and change: an investigation of the flaked stone artefacts from the Pleistocene deposits at Bone Cave southwest Tasmania, Australia* (Southern Forests Archaeological Report 3). Melbourne: Archaeology Program, School of Historical and European Studies, La Trobe University.

Holdaway, S. & N. Porch. 1995. Cyclical patterns in the Pleistocene human occupation of south-west Tasmania. *Archaeology in Oceania* 30: 74–82.

Jones, J. 1984. Hunters and history: a case study from western Tasmania, in C. Schrire (ed.) *Past and present in hunter gatherer studies*: 27–65. London: Academic Press.

Jones, J. 1990. From Kakadu to Kutikina: the southern continent at 18,000 years ago, in O. Soffer & C. Gamble (ed.) *The world at 18,000 BP. Volume 2: low latitudes*: 264–95. London: Unwin Hyman.

Kiernan, K., R. Jones & D. Ranson. 1983. New evidence from Fraser Cave for glacial age man in southwest Tasmania. *Nature* 301: 28–32.

Lambeck, K. 2004. Sea-level change through the last glacial cycle: geophysical, glaciological and palaeogeographic consequences. *C.R. Geoscience* 336: 677–89.

Lambeck, K. & E. Bard. 2000. Sea-level change along the French Mediterranean coast for the past 30 000 years. *Earth and Planetary Science Letters* 175: 203–22.

Langlais, M., S. Costamagno, V. Laroulandie, J-M. Pétillon, E. Discamps, J-B. Mallye, D. Cochard & D. Kuntz. 2012. The evolution of Magdalenian societies in south-west France between 18,000 and 14,000 calBP: changing environments, changing tool kits. *Quaternary International* 272–273 (12): 138–49.

Mackintosh, A.N., T.T. Barrows, E.A. Colhoun & L.K. Fifield. 2006. Exposure dating and glacial reconstruction at Mt. Field, Tasmania, Australia, identifies MIS 3 and MIS 2 glacial advances and climatic variability. *Journal of Quaternary Science* 21(4): 363–76.

Meiklejohn, C., G. Bosset & F. Valentin. 2010. Radiocarbon dating of Mesolithic human remains in France. *Mesolithic Miscellany* 21(1): 10–57.

Mellars, P. 2004. Reindeer specialization in the early Upper Palaeolithic: the evidence from south west France. *Journal of Archaeological Science* 31: 613–17.

Peregrine, P.N. 2004. Cross-cultural approaches in archaeology: comparative ethnology, comparative archaeology, and archaeoethnology. *Journal of Archaeological Research* 12(3): 281–309.

Pike-Tay, A., R. Cosgrove & J. Garvey. 2008. Archaeological evidence for seasonal land use patterns by late Pleistocene Tasmanian Aborigines. *Journal of Archaeological Science* 35: 2532–44.

Rasmussen, M. X. Guo, Y. Wang, K.E. Lohmueller, S. Rasmussen, A. Albrechtsen, L. Skotte, S. Lindgreen, M. Metspalu, T. Jombart, T. Kivisild, W. Zhai, A. Eriksson, A. Manica, L. Orlando, F.M. De La Vega, S. Tridico, E. Metspalu, K. Nielsen, M.C. Ávila-Arcos, J. Ví. Moreno-Mayar, C. Muller, J. Dortch, M.T.P. Gilbert, O. Lund, A. Wesolowska, M. Karmin, L.A. Weinert, B. Wang, J. Li, S. Tai, F. Xiao, T. Hanihara, G. Van Driem, A.R. Jha, F.X. Ricaut, P. De Knijff, A. B Migliano, I.G. Romero, K. Kristiansen, D.M. Lambert, S. Brunak, P. Forster, B. Brinkmann, O. Nehlich, M. Bunce, M. Richards, R. Gupta, C.D. Bustamante, A. Krogh, R.A. Foley, M.M. Lahr, F. Balloux, T. Sicheritz-Pontén, R. Villems, R. Nielsen, J. Wang, E. Willerslev. 2011. An Aboriginal Australian genome reveals separate human dispersals into Asia. *Science* 334: 94–8.

Shea, J. 2011. Refuting a myth about human origins. *American Scientist* 99: 128–35.

Shulmeister, J.D., T. Rodbell, M.K. Gagan & G.O. Seltzer. 2006. Inter-hemispheric linkages in climate change: paleo-perspectives for future climate change. *Climates of the Past* 2: 167–85.

Smith, M.A. & P. Hesse (ed.). 2005. *23°S: archaeology and environmental history of the southern deserts*. Canberra: National Museum of Australia Press.

Smith, M.A. & N.D. Sharpe. 1993. Pleistocene sites in Australia, New Guinea and island Melanesia: geographic and temporal structure of the archaeological record, in M.A. Smith, M. Spriggs & B. Fankhauser (ed.) *Sahul in review: Pleistocene archaeology in Australia, New Guinea and island Melanesia*: 37–59. Canberra: Department of Prehistory, Research School of Pacific Studies, Australian National University.

Soffer, O. 1987. *The Pleistocene Old World: regional perspectives*. New York: Plenum.

Stevens, R.E., R. Jacobi, M. Street, M. Germonprè, N. Conrad, S.C. Münzel & R.E.M. Hedges. 2008. Nitrogen isotope analyses of reindeer (*Rangifer tarandus*), 45,000 BP to 9,000 BP: palaeoenvironmental reconstructions. *Palaeogeography, Palaeoclimatology, Palaeoecology* 262: 32–45.

Taylor, K.C., G.W. Lamorey, G.A. Doyle, R.B. Alley, P.M. Grootes, P.A. Mayewski, J.W.C. White & L.K. Barlow. 1993. The 'flickering switch' of late Pleistocene climate change. *Nature* 361: 432–6.

Veth, P., M. Smith & P. Hiscock (ed.). 2005. *Desert peoples: archaeological perspectives*. Oxford: Blackwell.

Webb, C. & J. Allen. 1992. A functional analysis of bone tools from two sites in south west Tasmania. *Archaeology in Oceania* 25: 75–8.

Wobst, H.M. 1990. Afterword. Minitime and megaspace in the Palaeolithic at 18 K and otherwise, in O. Soffer & C. Gamble (ed.) *The world at 18,000 BP. Volume 2: low latitude*: 322–34. London: Unwin Hyman.

3 Strategies for Investigating Human Responses to Changes in Landscape and Climate at Lake Mungo in the Willandra Lakes, Southeast Australia

Nicola Stern, Jacqueline Tumney,
Kathryn E. Fitzsimmons and
Paul Kajewski

INTRODUCTION

The Willandra Lakes lie in the southwest corner of the Murray-Darling Basin on the edge of Australia's arid core (Figure 3.1). They were inscribed on the World Heritage register in 1981 partly because of their perceived potential for providing insights into the interplay between human and environmental history in a climatically sensitive area during the Late Pleistocene, a period of significant change in global climate (Mulvaney & Bowler 1981). This potential derives from the archaeological traces preserved in the alternating layers of sand and clay that record fluctuations in the amount and quality of water in the adjacent lakes, which were controlled primarily by changes in the amount of spring meltwater flowing westward from the Australian Alps.

This chapter reviews the features of the Willandra's sedimentary and archaeological records that demonstrate their extraordinary potential for generating information about past human activities and palaeoenvironments. Two studies illustrate how traces of past human activity can be linked directly to the palaeoenvironmental record, either through the stratigraphic context of the activity traces, which establishes their relationship to shifts in regional and global climates, or through their sedimentary context, which establishes the prevailing hydrological conditions.

The Commensurability of Records

A plethora of approaches have been and are being employed to explore the interface between environment, society and technology. Crosscutting these is a growing awareness of the need to consider the substantive implications of drawing inferences about the dynamics of past societies and changing environments from material remains preserved in the sedimentary record (Massey 1999; Holdaway et al. 2008). This is because the way in which

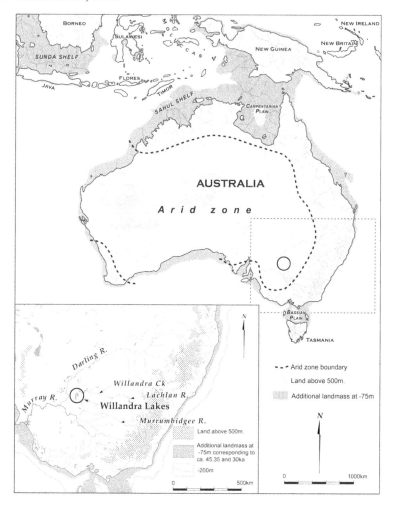

Figure 3.1 Map of Australia showing the main river systems draining the southeast Australian highlands, the location of the Willandra Lakes on the edge of the continent's arid core and the Willandra Creek, from which the lakes were periodically filled

material remains and their encasing sediments accumulate exerts a strong influence on the categories of behavioural and palaeoenvironmental information that can be generated from them (Binford 1981; Foley 1981; Bailey 1983, 2007; Stern 1993, 2008). Recent discussions have highlighted the frequent temporal mismatch between the archaeological data from which inferences about past behaviour are drawn and the palaeoenvironmental data used to investigate how past societies influenced, or were influenced by, changes in environment (Holdaway & Fanning 2010).

Such mismatches arise because both the geomorphic processes responsible for net sediment accumulation and the processes that result in the discard and accumulation of material remains vary in magnitude, frequency and locus of impact. As a result, the material remains and trace fossils incorporated into sediments do not necessarily accumulate in the same mode or at the same rate as each other, or as the sediments in which they are found (Behrensmeyer & Schindel 1983). An understanding of these relationships is needed in order to establish the commensurability of the behavioural and environmental records being studied (Holdaway & Fanning 2010; Holdaway et al. 2010). The depositional regime that operated in the Willandra Lakes during the Pleistocene preserved traces of discrete activities in sediments whose characteristics are a source of palaeoenvironmental information. This establishes a direct link between activities and the context in which they were undertaken and provides an opportunity to investigate human–environment interactions without having to grapple with the issues of commensurability posed by long-range correlations of data acquired from different depositional contexts (Smith et al. 2008; Holdaway et al. 2010).

CONTEXT

The Willandra Lakes

The Willandra Lakes are part of an extensive overflow system that is now dry but which filled periodically over the past 300,000 years at times when temperatures and evaporation were reduced sufficiently to result in a significant increase in the volume of spring meltwater flowing from the Australian Alps into the continent's semi-arid interior (Bowler 1971, 1998; Kemp & Rhodes 2010). During these times, the Lachlan River, which is one of three systems presently draining the western slopes of the southeast highlands, flowed into a now-abandoned channel known as the Willandra Creek. This flowed westward along the northern edge of the Riverine Plain until impeded by the Mallee dune fields; it then flowed south, filling several large and numerous smaller lake basins that lie nestled within the vast, arid plains formed by the linear and irregular dune fields (Bowler & Magee 1978).

    Fluctuations in lake hydrology were driven primarily by changes in the magnitude and frequency of flood pulses derived from the Australian Alps, which were linked to shifts in regional and global climates (Bowler 1971, 1998; Bowler et al. 2012). However, each of the lakes in this large, cascading overflow system has a unique hydrologic history, determined primarily by its position within the system (Figure 3.2). That history is recorded in the lunettes bounding the eastern margin of each lake, the sediments that accrued on their floors and the desert dunes that built up downwind. The lunettes have been the focus of efforts to write a palaeoenvironmental history of the Willandra partly because severe erosion of some has exposed their internal structure, facilitating investigation of their depositional history and partly because their alternating layers of

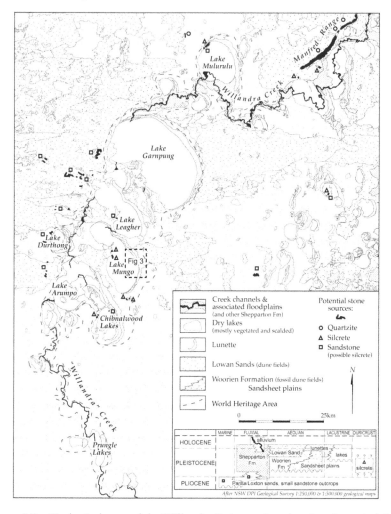

Figure 3.2   Geological map of the Willandra Lakes region showing the main lakes, the surrounding dune fields and the known locations of surface outcrops of silcrete, sandstone and quartzite

sand and clay reflect changes in the amount and quality of water in the adjacent lakes. Lake Mungo preserves a particularly detailed sedimentary record of past climate change because it filled via an overflow channel from Lake Leaghur and had no outflow, except evaporation (Bowler 1998: 148).

When the lakes were at overflow level, waves driven by the prevailing southwesterly winds washed sediments to their eastern margins and created high-energy beaches that accumulated quartz-rich sands and gravels, sometimes with cobble components. Sands blown from these beaches were deposited on top of the remnants of earlier dunes, creating vegetated fore-dunes with characteristically high-angle cross-set bedding (Bowler 1998; Bowler et al. 2012).

When the lakes fell below overflow level and evaporation exceeded precipitation, waters became increasingly saline and lake levels dropped, exposing part of the lake floor. On the exposed lake floor, salts precipitated from saline groundwater effloresced and broke up the sediment into sand-sized aggregates that were deflated by prevailing winds and dumped across the surrounding landscape. The resulting pelletal-clay lunette built up along the eastern margins of the sediment source, either on top of an existing quartz-sand lunette, or where a lake had contracted to depressions within its floor, along the eastern margins of those smaller basins. The pelletal-clay dunes exhibit low-angle, laminar bedding, and they sometimes contain components of reworked lake, beach or quartz-dune sands (Bowler 1973, 1998).

Thus, clean quartz-sand and gravel or cobble beaches reflect lake-full conditions, while pelletal clays (sometimes containing components of sand) represent low and fluctuating lake levels. Soils formed during periods of relative stability, for instance, when the lakes had just refilled and there was a temporary cutoff in sediment supply or when the lakes were completely dry and the landscape was vegetated (Bowler 1998).

Stratigraphy of the Mungo Lunette

Bowler (1998) described five stratigraphic units representing a succession of lake transgressions separated by periods of landscape stability. The low-ermost is the Golgol unit, which represents an early phase of lake activity that was followed by a long period of landscape stability (Bowler 1998: 137, 140). Approximately 55,000 years ago the lakes refilled, ushering in a long period of sustained freshwater conditions that were associated with the formation of a quartz-sand lunette. This is the Lower Mungo unit, which contains some of the oldest best-dated archaeological traces on the continent (Bowler et al. 1970; Bowler et al. 2003; Bowler 1998: 137–40). The bed of dark-grey pelletal clay that overlies these sands reflects a shift towards more arid conditions *c.* 40,000 years ago and marks the base of the Upper Mungo unit. The alternating layers of pelletal clay and clean quartz sands of the Upper Mungo unit reflect regularly oscillating lake levels, while the overlying Arumpo unit represents intermittent lake-full conditions, with short-lived episodes of drying and deflation. The Zanci unit is the youngest of those described by Bowler, and its laminar sands record the final drying of the lake (Bowler 1998: 139–40).

It is salutary to note that Bowler's stratigraphic summary is based on detailed study of a few key sections at the southern end of the Mungo lunette and a major gully section through the Outer Arumpo lunette that provides a rare window into the associated lake margin and lake deposits.

These sections were recorded with the aim of building a picture of regional palaeoenvironments, rather than a detailed hydrologic history of the lakes. Each of the stratigraphic units that Bowler describes incorporates numerous depositional and erosional events whose impact varied along the length of the lunette depending on the scale of the event, prevailing topography and wind patterns. For this reason analysis and interpretation of the palaeoenvironmental context of archaeological traces have to be based on detailed records of the stratigraphic and sedimentary context of the specific features or assemblages under investigation.

Traces of Past Human Activity

Archaeological traces were incorporated into the lunette sediments as they accumulated and are continually being exposed on the modern surface through ongoing erosion. The most visible of these are the scatters of hearthstones, stone artefacts and fragmented animal bones that are strewn across the surfaces of the exposures, because their encasing sediment has been removed and/or because they have been transported down the rill and gully systems and concentrated by waning flow. It is sometimes possible to determine the stratigraphic unit from which these lag and transported assemblages have eroded (Tumney 2011). However, this requires detailed analysis of the distribution of debris in relation to stratigraphic boundaries, palaeotopography and modern topographic features, because the redistribution and mixing of material is the outcome of a myriad of interacting processes and visual inspection is rarely sufficient to establish their impact. In some topographic settings, like the low exposures at the front of the lunette, surface assemblages could be derived from any or all of the sediments that have accumulated since humans first settled the area.

The less visible component of the archaeological record comprises thousands of features and isolated finds that are partially embedded in sediment and have precise stratigraphic and palaeotopographic provenience. It also includes discrete clusters of surface debris that exhibit no evidence for post-exposure weathering or fragmentation, indicating recent removal of encasing sediments, because once exposed, these features disintegrate and disperse within one to two years. These activity traces include a variety of cooking hearths, clusters of burned and unburned food remains, stone tools and debris from their manufacture and repair, as well as ochre pellets, unworked silcrete cobbles, large cores, grindstones and other tools. Most of these arguably result from single episodes of activity: the lighting of a fire to cook a small marsupial, the striking of a few flakes from a core or the cooking of an emu egg. Although they conjure up evocative images of past lives, they nevertheless present archaeologists with all the challenges inherent in studying 'landscape palimpsests' (Bailey 1983, 2007; Stern 1993). Few of these activity traces have been documented or studied in any detail (but see Bowler et al. 1970; Allen 1972), reflecting the difficulty of studying archaeological remains scattered through vast landforms that have complex depositional and erosional histories, but which are also part of an active landscape.

SETTLEMENT HISTORY

It is possible to identify the palaeoenvironmental conditions that brought people to the margins of the Willandra Lakes more frequently and/or for longer duration and/or in larger numbers by investigating the distribution of archaeological traces in the Mungo lunette in relation to stratigraphic and/or sedimentary units that represent different lake conditions. The data discussed here were recorded during the first systematic foot survey of the Mungo lunette ever attempted and are part of a broader investigation into the changing pattern of life in the Willandra over the past 45,000 years.

Integration of archaeological and geological data is fundamental to this endeavor and is achieved by recording the locations of cultural features and stratigraphic boundaries using the same three-dimensional grid system that is tied into the Geocentric Datum Australia. During the foot survey, records are made of both the locations of past activity traces and the sediment in which they are embedded. Stratigraphic boundaries are traced out across the exposures and recorded on geo-rectified digital air photos captured in 2007; these overlays are subsequently digitized. To establish the geological context of the archaeological features, mapping is undertaken in much finer detail than would be required simply to establish the lake's depositional and palaeoenvironmental history.

This discussion focuses on a stretch of lunette extending from the 'Walls of China tourist site' approximately 770m northwards along a wave-cut bench formed on the Golgol unit during a high lake stand some time after the lakes refilled *c.* 55,000 years ago (Figure 3.3). Nine stratigraphic units are found in this area, six of which are ancient lunette deposits and three of which are active depositional environments resulting from the reworking of ancient lunette sediments (see Table 3.1). This sequence is similar, but not identical, to the sequence described by Bowler (1998; Bowler et al. 2003) for the southern end of the Mungo lunette.

In both areas, the core of the lunette is made up of sediments deposited during an earlier phase of lake activity (Unit A = Golgol Unit). However, in the central portion of the lunette, sediments that were deposited during the initial phases of the current lake cycle (Unit B = Lower Mungo and Unit C = Upper Mungo) are thin and laterally discontinuous, reflecting limited deposition in this area (perhaps as a result of more southerly sediment trans-port) as well as the impact of recent, severe erosion. At the southern end of the lunette, the maximum vertical and lateral thickness of these units is much greater (Bowler 1998: fig. 2).

The greatest volume of sediment in the central portion of the Mungo lunette consists of predominantly sandy sediments that accumulated between about 25,000 and 15,000 years ago (Unit E). Across much of the

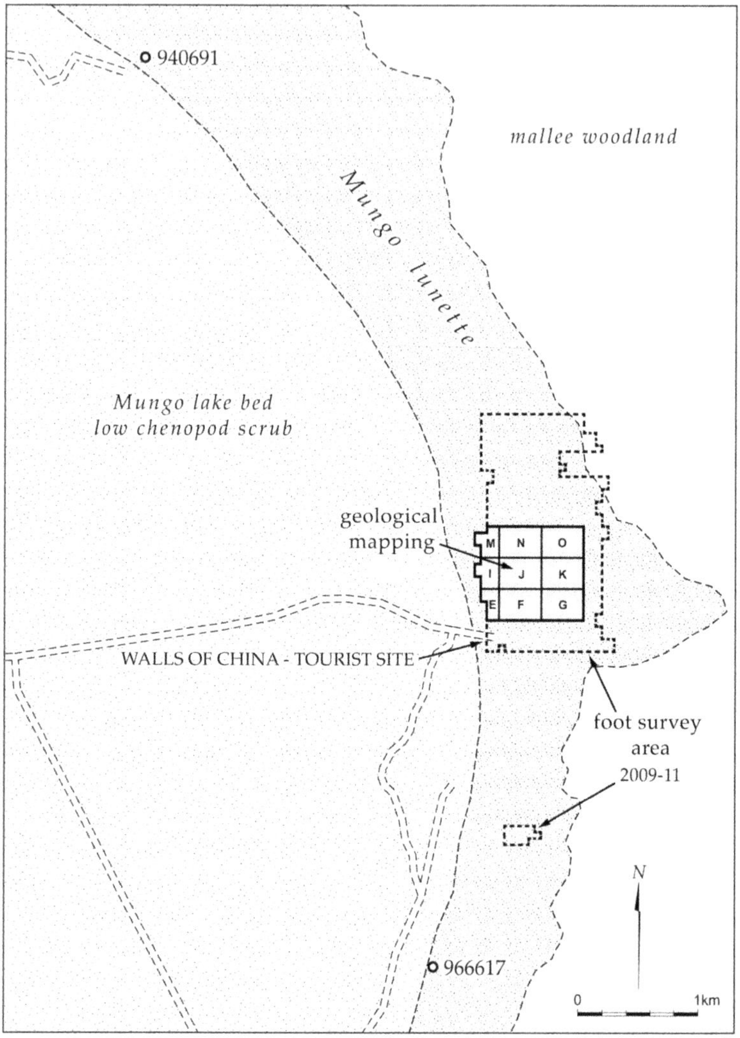

Figure 3.3   Map of part of the northern Mungo lunette, showing the location of the study areas referred to in the text

study area, these sands bank up against, and overlie, the Golgol wave-cut bench. The sands at the base of this unit are massive, but they are overlain by a thick sequence of alternating quartz sands, sandy clays and pelletal clays. Discontinuous and weakly developed soils occur at various levels within Unit E, but this sedimentary package does not contain a laterally extensive or well-developed soil horizon that could serve as a marker horizon. Thus, it is the lateral equivalent of the Arumpo and Zanci units defined by Bowler (1998).

Table 3.1 Archaeological traces, their stratigraphic, sedimentary and palaeoenvironmental context and approximate age. The stratigraphic units in the study are correlated with those described by Bowler (1998) and Bowler et al. (2003, 2012). Age estimates include data from Bowler (1998), Bowler et al. (2003, 2012), Gillespie (1998), Bowler & Price (1998) and Dare-Edwards (1979)

Stratigraphic unit

This study	Bowler 1998	Description	Inferred environment	Approximate age	Archaeological traces
J	–	Pale clayey sands filling active gullies on the lakeward flanks of the lunette and building up on outwash fans at its base	Lake dry; reworking of sediments through rain action	Present	Some archaeological material recemented into modern gully-fills and outwash fans
H	–	Well-sorted, fine sands building up on the crest and lee flanks of the lunette	Lake dry; reworking of sediments through wind action	Present	None
I	–	Brown, clayey sands making up alluvial fans on the lakeward flank of the lunette, resulting from fluvial reworking of lunette sediments	Lake dry; locally more humid conditions	Mid-Holocene–Present	Low density; predominantly heat-retainer hearths and clusters of stone artefacts
F	–	Pale, unconsolidated sands on the crest and leeward flanks of the lunette, built up through reworking of older lunette sediments. Multiple, discontinuous brown sandy soils have formed at various levels within this unit	Lake dry; locally more arid conditions with short-lived episodes of landscape stability	~14–~8ka	Low density; predominantly heat-retainer hearths and clusters of stone artefacts (many with refits)

(Continued)

Table 3.1—(Continued)

This study	Bowler 1998	Description	Inferred environment	Approximate age	Archaeological traces
Stratigraphic unit					
E	Arumpo/ Zanci	A predominantly sandy unit with massive, quartz sands at its base that are overlain by alternating layers of pale quartz sands and pelletal clays and laminar clayey sands. Discontinuous, weakly developed soils formed on different beds throughout the sequence	A phase of sustained lake-full conditions followed by oscillating lake levels; the lake may have dried out occasionally, but most of the time retained a substantial body of water	~25–15ka	Archaeological traces abundant and varied; includes a variety of heat-retainer and baked sediment hearths, many with associated food remains; fishbone hearths occur in restricted portions of the sequence, eggshell hearths more widespread; clusters of stone artefacts (including refits); isolated finds include shell tools and grindstones
D	–	Thin, discontinuous, steeply dipping red sandy unit, with beach pebbles on the lakeward flank	Sustained high lake level	–	Archaeological traces moderately abundant, but sample size is small because of limited area of exposure
C	Upper Mungo	Thin unit comprising alternating pale sands, clayey sands and pelletal clays exposed in small patches in the southern part of the study area. Capped by a well-developed soil	Oscillating lake levels	~40–~30ka	Greatest density and diversity of archaeological traces despite small area of exposure, includes a variety of heat-retainer and baked sediment hearths containing terrestrial and/or lacustrine resources

B	Lower Mungo	Thin, discontinuous, well-sorted red quartz sands grading up-section to pale quartz sands, with freshwater shell fragments	Sustained high lake levels	~55–40ka	Archaeological traces moderately abundant, includes hearths with highly fragmented and sometimes burned food remains
A	Golgol	Red, indurated sands with a well-developed palaeosol	High lake levels followed by drying of the lake and intensive soil formation. A wave-cut bench formed during a subsequent high lake level is a prominent feature of the study area	>130ka	None

Sediments continued to accumulate on the Mungo lunette after the lake dried out *c.* 15,000 years ago in response to local palaeoenvironmental conditions, rather than those in a distant catchment. The sediments at the top of the lunette sequence represent aeolian reworking of ancient sediments under relatively more arid conditions (Unit F) as well as fluvial reworking of sediments under relatively more humid conditions (Unit I).

Sediments continue to build up on the Mungo lunette as a result of both fluvial and aeolian reworking of older deposits. Water washes sediments down the expanding rill and gully systems, contributing to the build-up of alluvial fans at the base of the lunette and to the draping of sheetwash deposits across the lower slopes of the lakeward flanks of the lunette (Unit J). When conditions are dry, prevailing winds pick up the sands and redeposit them on the mobile dunes that have formed on the leeward side of the lunette (Unit H). These active depositional environments cover approximately 40% of the current study area.

Traces of past human activity are not distributed evenly amongst the stratigraphic units just described (Table 3.1). Significantly more features are preserved in sediments representing oscillating lake levels than predicted on the basis of their area of exposure, whether the total number of sites or the number of individual features (heat-retainer hearths, baked sediment hearths, isolated *in situ* finds, stone artefact clusters and animal bone clusters) is considered. In these alternating layers of sand and clay, 50% of the hearths are embedded in clean quartz sands that accumulated when lake levels were high, and the other 50% are embedded in pelletal clays that formed when the lake floor was partly exposed and there was discharge of saline groundwater. This indicates that people were attracted to the lake's margin by the conditions created by oscillating lake levels, rather than short-lived stands of high or low lake levels.

Archaeological traces are more abundant than predicted on the basis of area of exposure in sediments representing sustained high lake levels (Unit B and the massive sands at the base of Unit E). However, they are less abundant than predicted in the reactivated sands at the top of the lunette sequence (Unit H) and in the alluvial fans that have been building up at the toe of the lunette since the mid-Holocene (Unit I).

The distribution of archaeological traces through the stratigraphic sequence does not support the long-standing popular perception that people were attracted to the Willandra Lakes when they were full of freshwater; nor does it support long-standing speculation that aquatic resources were the focus of subsistence activities (Allen 1998: 212; Mulvaney & Kamminga 1999: 197). Bowler (1998: 147) suggested that people fell back to the margins of the Willandra Lakes when the hinterland dried out. When lake levels were high, surface water would have been abundant on the surrounding plains, removing a critical constraint on how far, how often, and for how long people could go there to forage for food or meet up with other social groups. Frequent flood pulses would have enhanced the biological productivity of the overflow system, in the same way that they enhance the productivity of extant wetlands in the Murray-Darling Basin (Robertson et al. 1999; Scholz et al. 2002; King et al. 2009). The lake and the lake margin zone would have supported a greater diversity of potential food items, in greater abundance and with greater predictability when flood pulses regularly recharged the system than during periods of sustained lake-full conditions. Thus, it is not surprising that more activity traces were incorporated into the accumulating lunette when lake levels were low and oscillating than when they were high and stable.

Hearths containing *either* terrestrial *or* lacustrine resources are both more abundant in sediments representing oscillating lake levels than predicted. However, hearths containing *both* types of resources are especially so. This suggests that when Lake Mungo was an important focus of people's lives, foods were being acquired from both the lake and the surrounding plain. Fishbone hearths are a tiny proportion of the entire hearth sample, but are restricted to laminar sands that accumulated rapidly under oscillating lake conditions. No shell middens and only two small scatters of non–culturally accumulated shell were recorded in the present study area. Despite Johnston's (1993) argument that the importance of aquatic resources has been overestimated, characterisation of the Willandra's archaeological record as one of middens and stone tools persists (for example, Holdaway & Fanning 2010: 73).

A recent compilation of radiocarbon dates on freshwater mussels and fish otoliths from across the Willandra Lakes shows that they cluster in two time intervals, one corresponding to the Last Glacial Maximum (about 20,500 to 17,000 years ago) and one to the period between about 40,000 and 36,500 years ago when there was a marked trend towards regional aridity (Bowler et al. 2012). This suggests that fish and shellfish were important fallback foods. Although it is sometimes suggested (for example, Allen & Holdaway 2009: 99) that the Willandra was so lacking in surface water and so windy and dusty during the Last Glacial Maximum that people abandoned the region, both the dates on aquatic resources (Bowler et al. 2012) and the formation of a predominantly sandy lunette at Lake Mungo at that time indicate that this was not the case.

Earlier research in the Willandra suggested that when the lakes dried out at the end of the Last Glacial Maximum the focus of people's lives shifted to the Darling River, and the now-vegetated overflow system became part of the hinterland, visited intermittently (Allen 1974). However, the hearths and stone artefacts found on the lake floor (Johnston & Clark 1998), in the reactivated dune sediments at the top of the lunette and in the alluvial fans at its base show that people lived in this landscape under varied palaeoenvironmental conditions after the lakes had dried out. The restricted range of activity traces preserved in these sediments is consistent with the suggestion that people were highly mobile, shifting the locus of foraging and other activities in response to rainfall and intermittent and localised abundance of key resources. Identifying the resources exploited under different conditions will require detailed investigation of the hearths and food remains preserved in these sediments.

TECHNOLOGICAL CHANGE AND PALAEOENVIRONMENTS

Two shallow erosion blowouts in the northern part of the Mungo lunette (Figure 3.3) provide a case study for exploring the technological strategies employed at different times under correspondingly different palaeoenvironmental conditions. Detailed analysis of the distribution of surface stone artefacts in relation to stratigraphic and topographic boundaries establishes the stratigraphic origin of these assemblages (Tumney 2011: 114–42). At blowout 940691 the artefacts are associated with one of two sedimentary units (Table 3.2). The older (OS) unit comprises massive, pale quartz sands associated with a period of consistently high lake levels. The upper (GC) unit consists of layers of alternating clays and clayey sands of varying thickness, representing a shift to fluctuating lake levels. Optical luminescence age estimates indicate that the entire sediment package was deposited between about 24,000 and 20,000 years ago (Tumney 2011: 73–8).

At blowout 966617, a similar sequence of sediments has been identified (Table 3.2), but here the sandy and clayey units are separated by a layer of reworked carbonate nodules and rhizomorphs derived from the main body of the lunette, which represent a major erosional event (Barrows, *pers. comm.*). The clayey sediments of the upper unit are of a similar age to those in the northern blowout, indicating ephemeral lake conditions between about 22,000 and 16,000 years ago (Tumney 2011: 88–93).

The chipped stone artefacts in these assemblages were made predominantly from silcrete, a silicified sandstone available from several locations within 20km of the Mungo lunette (Figure 3.2). The grain size and quality of these silcretes varies considerably but can be grouped into coarse-grained

Table 3.2 Palaeoenvironmental context and approximate age of each artefact assemblage

Artefact assemblage	Sediment description	Inferred environment	Approximate age (ka)
966617	Alternating sands and clays	Fluctuating lake levels, drying	<22 – <16
GC, 940691	Alternating sands and clays	Fluctuating lake levels, drying	21 – <20
OS, 940691	Massive sands	Sustained high water	24 – 21

and fine-grained varieties. Past studies have suggested that even the fine-grained silcrete is of low flaking quality (Mühlen-Schulte 1985; Webb & Domanski 2008). The silcrete sources that have been inspected all contain predominantly fine-grained silcrete; the source of the coarser-grained variety is currently unknown.

The third raw material present in significant quantities is a high-quality quartzite, the nearest known source of which lies 70km to the northwest in the Manfred Range (Figure 3.2). The presence of artefacts made from three raw materials of varying quality that could be acquired at varying distances from the lunette provides an opportunity to investigate the impact of raw material quality, transport distance and environmental context on patterns of use and discard.

The characteristics of these artefact assemblages provide insights into the relative mobility of social groups—that is, differences in the frequency and/or distance and/or duration of moves from one area to another. The more mobile a group, the more likely they are to *encounter* a particular raw material (Shiner 2008: 33). Whether they then *collect* material depends on its quality and the likelihood of encountering higher-quality stone (Bird & O'Connell 2006: 147). Given the poor quality of silcrete in this region it is highly likely that quartzite was collected whenever it was encountered. Thus, the quantity of quartzite in each assemblage, and the intensity with which it was worked, reflects the relative probability that a group regularly travelled some 70km to acquire it.

The smaller average artefact size, higher flake:core ratio, greater frequency of retouch and higher proportion of tools with 'formal' (i.e. relatively regular) retouch indicate that quartzite was worked more intensively than silcrete, as predicted by its higher quality and greater procurement cost.

In comparison with the artefacts made from fine-grained silcrete and quartzite, those made from coarse-grained silcrete are less abundant, larger and not worked or retouched as intensively. This pattern is likely to reflect the poor flaking quality of this material.

Coarse-grained silcrete artefacts are smallest in the 966617 assemblages, and coarse-grained silcrete is relatively more abundant (by mass) in the 940691 GC assemblage. If raw material quality is not considered, this could indicate a source that is closer to 940691, at least during the lower-water phase. However, the poor flaking quality of coarse-grained silcrete suggests an alternative interpretation. Low-quality materials are only likely to be utilised if the probability of encountering better-quality material is low. Thus, the greater quantity of coarse-grained silcrete in the GC assemblage may reflect a reduced ability to obtain sufficient quantities of either fine-grained silcrete or quartzite.

Artefacts in the GC assemblage that are made from fine-grained silcrete and quartzite exhibit more retouch, and more of them can be classified as 'formal' tool types than in the other assemblages. The size of the quartzite artefacts, and the way in which the fine-grained silcrete cores were rotated and discarded, also indicates greater concern for the conservation of these materials in this assemblage. Together with the evidence from the coarse-grained silcrete artefacts, these patterns are interpreted as the result of lower levels of mobility.

Conversely, the greatest degree of mobility is inferred from the 966617 assemblage. Quartzite is most abundant in this assemblage, and quartzite artefacts are relatively large compared with those in the other assemblages. Formal retouch on both quartzite and fine-grained silcrete artefacts is less frequent in this assemblage, and patterns of core rotation and discard indicate minimal concern for the conservation of fine-grained silcrete.

Evidence for the differential removal of quartzite flakes from the 940691 OS assemblage makes this assemblage more difficult to interpret. However, it is more similar to the 966617 assemblage than the GC assemblage, suggesting high levels of relative mobility.

The pattern of highest mobility in the 966617 assemblage and lowest mobility in the GC assemblage is surprising, given that the sedimentary evidence indicates that these assemblages accumulated during similar environmental conditions, while the OS assemblage represents quite different conditions. One possible interpretation is that the shift from high to fluctuating water levels made the lake margins a more attractive place to forage, whilst continued drying reduced their productivity, resulting in more mobile foraging strategies in which the lakes were probably peripheral. This interpretation is consistent with the foot survey data, which show that greater quantities of debris accumulated in the lunette when the levels in Lake Mungo were oscillating than when it was either completely full or completely dry.

DISCUSSION

These studies show that the long-perceived potential of the Willandra for providing insights into human responses to changes in landscape and climate has not been misplaced. However, to ensure commensurability of behavioural and palaeoenvironmental information, tightly integrated archaeological and geological data are needed. As these studies illustrate, this can be achieved by describing the sedimentary context of all features and finds and through detailed mapping of sediments and/or stratigraphy and archaeological finds using the same three-dimensional grid system. Sedimentary and/or stratigraphic units representing correspondingly different palaeoenvironmental conditions are the analytical units from which interpretations are then developed.

This methodology differs in its substantive implications from earlier efforts to investigate human–environment interactions in the Willandra, which involved extrapolating palaeoenvironmental information generated from a few well-studied sections to other parts of the system and treating geographically and temporally scattered archaeological traces as material fallout from an ethno-historically recorded behavioural system (for critical re-evaluations see Allen 1998; Allen & Holdaway 2009). It was this approach that led Holdaway & Fanning (2010: 73) to argue that there is a serious mismatch in scale between the behavioural and palaeoenvironmental records in the Willandra. However, this apparent mismatch is an artefact of methodology and not an inherent property of the Willandra's archaeological and geological records.

Systematic documentation of the archaeological traces preserved in different strata and sediments along a stretch of the northern Mungo lunette shows that traces of people's activities accumulated in this landscape under varied conditions: when it was dominated by a stable, freshwater lake, when the lake fell below overflow and its levels fluctuated dramatically, while the lake was drying and after it had dried out completely. Given the magnitude of the landscape and ecological changes associated with these changes in lake hydrology, it is reasonable to surmise that there were concomitant changes in diet and food-getting strategies, patterns of movement across the landscape, the way technological systems were organised and the extent of the social networks that people maintained.

The data reviewed here provide some insights into aspects of these changing behaviours. The abundance and diversity of debris in sediments indicative of regular fluctuations from fresh to saline conditions undoubtedly reflect the greater biological productivity associated with flood pulses and suggest that during these times, the lake margin was an attractive base from which to forage for food. Subtle shifts in stone technology point to a reduction in mobility, suggesting that more time was, indeed, spent along the lake's margin. Lacustrine resources were incorporated into the record more often whilst these conditions prevailed, but they did not replace terrestrial prey. More detailed information about diet will require analysis of the food remains recovered from dozens of intact hearths preserved in different sedimentary contexts, work that is just beginning.

These two studies highlight the importance of not assuming that the relationship between environmental change and shifts in behaviour will be straightforward. Changes in diet may have involved shifts in the relative importance of lake versus terrestrial resources and/or shifts in the ranking of terrestrial prey species, many of which have broad ecological tolerances and would not have vanished from the landscape even when the lake dried out. Similarly, the latitude in form and edge characteristics that many stone tools can exhibit and still perform the same task effectively means that changes in technology may have involved adjustments in the relative importance of different raw material sources and the intensity with which different materials were worked, rather than the appearance of new tools or tool kits.

Identifying these subtle changes also requires an improved understanding of the relationship between the food remains at individual hearths and the dietary information that can be inferred when the contents of many hearths are aggregated to create a meaningful unit for palaeoenviromental analysis.

Because the record contains hundreds of discrete activity traces, it presents a remarkable opportunity to explore those relationships and, in so doing, enhance current understanding of the interplay between human activities and past environments.

ACKNOWLEDGMENTS

This research began with permission from the Elders' Council of the Three Traditional Groups and the Technical and Scientific Advisory Committee of the Willandra Lakes Region World Heritage Area (WLRWHA) and continued under the stewardship of the Elders' Council of the Two Traditional Tribal Groups of the WLRWHA. It is a privilege to have been welcomed into Paakantyi, Ngiyampaa and Mutthi Mutthi country, and we are grateful to the Elders for their willingness to discuss the aims and scope of this work, their contributions to the fieldwork and their ongoing support of our endeavors. This research was funded by an Australian Research Council Linkage grant (LP0775058) and an Australian Research Council Discovery grant (DP1092966) and supported by La Trobe University, the Australian National University and the Max Planck Institute for Human Evolutionary Studies. We thank our industry partners, the WLRWHA Elders' Council and the NSW Department of Environment and Heritage, and our colleagues for their contributions. We are indebted to the project's Cultural Heritage Officer, Daryl Pappin, for his dedicated assistance with all aspects of the field research and to the La Trobe University undergraduate students and Mungo National Park Discovery Rangers for the commitment and enthusiasm they brought to the fieldwork. In his retirement Rudy Frank has been extraordinarily generous with his time and skills, assisting with the fieldwork and preparing the figures.

REFERENCES

Allen, H. 1972. Where the crow flies backwards: man and land in the Darling Basin. Unpublished PhD dissertation, Australian National University.

Allen, H. 1974. The Bagundji of the Darling Basin: cereal gatherers in an uncertain environment. *World Archaeology* 5: 309–22.

Allen, H. 1998. Reinterpreting the 1969–1972 Willandra Lakes archaeological surveys. *Archaeology in Oceania* 33: 207–20.

Allen, H. & S. Holdaway. 2009. The archaeology of Mungo and the Willandra Lakes: looking back, looking forward. *Archaeology in Oceania* 44: 96–106.

Bailey, G.N. 1983. Concepts of time in Quaternary prehistory. *Annual Review of Anthropology* 12: 165–92.

Bailey, G.N. 2007. Time perspectives, palimpsests and the archaeology of time. *Journal of Anthropological Archaeology* 26: 198–223.

Behrensmeyer, A.K. & D. Schindel. 1983. Resolving time in paleobiology. *Paleobiology* 9: 1–8.

Binford, L.R. 1981. Behavioural archeology and the 'Pompeii premise.' *Journal of Anthropological Research* 37: 195–208.

Bird, D. & J.F. O'Connell. 2006. Behavioral ecology and archaeology. *Journal of Archaeological Research* 14: 143–88.

Bowler, J.M. 1971. Pleistocene salinities and climatic change: evidence from lakes and lunettes in southeastern Australia, in D.J. Mulvaney & J. Golson (ed.) *Aboriginal man and environment in Australia*: 47–65. Canberra: Australian National University Press.

Bowler, J.M. 1973. Clay dunes: their occurrence, formation and environmental significance. *Earth Science Reviews* 9: 315–38.

Bowler, J.M. 1998. Willandra Lakes revisited: environmental framework for human occupation. *Archaeology in Oceania* 33: 120–55.

Bowler, J.M., R. Gillespie, H. Johnston & K. Boljkovac. 2012. Wind v water: glacial maximum records from the Willandra Lakes, in S.G. Haberle & B. David (ed.) *Peopled landscapes: archaeological and biographic approaches to landscapes*: 271–96. Terra Australis 34. Canberra: Australian National University.

Bowler, J.M., H. Johnston, J.M. Olley, J.R. Prescott, R.G. Roberts, W. Shawcross & N.A. Spooner. 2003. New ages for human occupation and climatic change at Lake Mungo, Australia. *Nature* 421: 837–40.

Bowler, J.M., R. Jones, H. Allen & A.G. Thorne. 1970. Pleistocene human remains from Australia: a living site and human cremation from Lake Mungo, western New South Wales. *World Archaeology* 2: 39–60.

Bowler, J.M. & J. Magee. 1978. Geomorphology of the Mallee region in semi-arid northern Victoria and western New South Wales. *Proceedings of the Royal Society of Victoria* 90: 5–25.

Bowler, J.M. & D.M. Price. 1998. Luminescence dates and stratigraphic analyses at Lake Mungo: review and new perspectives. *Archaeology in Oceania* 33: 156–68.

Dare-Edwards, A.J. 1979. *Late Quaternary Soils on Clay Dunes of the Willandra Lakes*, Unpublished PhD dissertation, Australian National University.

Foley, R. 1981. A model of regional archaeological structure. *Proceedings of the Prehistoric Society* 47: 1–7.

Gillespie, R. 1998. Alternative timescales: a critical review of Willandra Lakes dating. *Archaeology in Oceania* 33: 169–82.

Holdaway, S.J. & P. Fanning. 2010. Geoarchaeology in Australia: understanding human–environment interactions, in P. Bishop & B. Pillans (ed.) *Australian landscapes*. Geological Society of London, Special Publications 346: 71–85.

Holdaway, S.J., P. Fanning & E.J. Rhodes. 2008. Assemblage accumulation as a time-dependent process in the arid zone of western New South Wales, Australia, in S. Holdaway, & L. Wandsnider (ed.) *Time in archaeology: time perspectivism revisited*: 110–33. Salt Lake City: The University of Utah Press.

Holdaway, S.J., P. Fanning, E.J. Rhode, S.K. Marx, B. Floyd & M.J. Douglass. 2010. Human response to palaeoenvironmental change and the question of temporal scale. *Palaeogeography, Palaeoclimatoloty, Palaeoecology* 292: 192–200.

Johnston, H. 1993. Pleistocene shell middens of the Willandra Lakes, in M.A. Smith, M. Spriggs & B. Fankhauser (ed.) *Sahul in review: Pleistocene archaeology in Australia, New Guinea and island Melanesia*: 197–203. Department of Prehistory, Research School of Pacific Studies. Canberra: Australian National University.

Johnston, H. & P. Clark. 1998. Willandra Lakes archaeological investigations 1968–98. *Archaeology in Oceania* 33: 105–19.

Kemp, J. & E. Rhodes. 2010. Episodic fluvial activity of inland rivers in southeastern Australia: palaeochannel systems and terraces of the Lachlan River. *Quaternary Science Reviews* 29: 732–52.

King, A.J., Z. Tonkin & J. Mahoney. 2009. Environmental flow enhances native fish spawning and recruitment in the Murray River, Australia. *River Research and Applications* 25: 1205–18.

Massey, D. 1999. Space–time 'science' and the relationship between physical geography and human geography. *Transactions of the British Institute of Geography* NS 24: 261–76.

Mühlen–Schulte, R. 1985. Mungo rocks. A technological analysis of stone assemblages from Lake Mungo. Unpublished BA honours thesis, Australian National University.

Mulvaney, D.J. & J.M. Bowler. 1981. Lake Mungo and the Willandra Lakes, in *The heritage of Australia: the illustrated register of the national estate*: 180–83. Sydney: Macmillan.

Mulvaney, D.J. & J. Kamminga. 1999. *Prehistory of Australia*. Sydney: Allen and Unwin.

Robertson, A.I., S.E. Bunn, F. Walker & I. Boon. 1999. Sources, sinks and transformations of organic carbon in Australia flood plain rivers. *Marine and Freshwater Research* 50: 813–29.

Scholz, O., B. Gawne, B. Ebner & I. Ellis. 2002. The effects of drying and re-flooding on nutrient availability in ephemeral deflation basin lakes in western New South Wales. *River Research and its Applications* 18: 185–96.

Shiner, J. 2008. *Place as occupational histories: an investigation of the deflated surface archaeological record of Pine Point and Langwell Stations, Western New South Wales, Australia*. BAR International Series 1763. Oxford: Archaeopress.

Smith, M.A., A.N. Williams, C.S.M. Turney & M.I. Cupper. 2008. Human-environment interactions in Australian drylands: exploratory time-series analysis of archaeological records. *The Holocene* 18: 389–401.

Stern, N. 1993. The structure of the Lower Pleistocene archaeological record: a case study from the Koobi Fora Formation in northwest Kenya. *Current Anthropology* 34: 201–25.

Stern, N. 2008. Stratigraphy, depositional environments and palaeolandscape reconstruction in landscape archaeology, in B. David & J. Thomas (ed.) *Handbook of landscape archaeology*: 365–78. Walnut Creek: Left Coast Press.

Tumney, J. 2011. Environment, landscape and stone technology at Lake Mungo, southwest New South Wales, Australia. Unpublished PhD dissertation, La Trobe University.

Webb, J. & M. Domanski. 2008. The relationship between lithology, flaking properties and artefact manufacture for Australian silcrete. *Archaeometry* 50: 555–75.

4 A New Ecological Framework for Understanding Human–Environment Interactions in Arid Australia

Simon J. Holdaway, Matthew J. Douglass and Patricia C. Fanning

INTRODUCTION

Twenty years ago Stafford Smith and Morton (1990) published an ecological assessment of Australia as a series of propositions that illustrated why the continent stood apart ecologically from other arid areas around the world. The assessment has now been updated in light of recent work (Morton et al. 2011). Both studies provide an opportunity to question how archaeologists working in Australia have utilised ecological information: to what degree have they developed understandings of hunter–gatherer behaviour in the deserts that are different from similar studies in other regions of the world? To answer this question the ecological assessments are summarised, followed by a review of how archaeologists have considered the Australian environment. A case study from recent work in western New South Wales is provided to illustrate how the propositions put forward in the ecological assessment can be investigated archaeologically.

People do not necessarily behave in concert with their environment; hence archaeologists have replaced a deterministic model for human–environment interactions with an understanding of how people are both affected by and affect the environment (Kirch 2005). For hunter–gatherers, the theoretical approaches archaeologists use to investigate this interdependence originated from studies of contemporary groups living in places outside Australia. Binford (1979), for instance, worked with high latitude Nunamuit hunters, while Yellen (1977) undertook ethnoarchaeological work with the !Kung in the Kalahari. Other ethnoarchaeological work was carried out in Australia (e.g. Hayden 1979; O'Connell & Hawkes 1981, 1984; Gould 1991). In all these cases, ethnographic work was limited in duration, spanning a few months through to one to two years. This placed some limitations on the value of the studies for understanding longer term interaction between people and their environment.

In an often cited study, Binford (1980) proposed a dichotomy between foragers and collectors as a way of understanding the nature of hunter–gatherer organisation in the ethnographic case studies he was considering.

Binford's intent was to show the relationship between the socioeconomy of the Nunamuit and their material culture use. 'Foragers' and 'collectors' were devised as conceptual models useful for explaining how different material records might be generated in different situations. However, upon publication, the models were quickly adopted as direct explanations for artefact assemblages and site types in many different places and periods. If, as the recent ecological assessments suggest, Australia has a number of differences from other areas of the world that supported hunter–gatherer populations, it remains an unanswered question as to whether models derived from outside Australia are useful for describing and explaining Australian hunter–gatherer behaviour. Here, a set of recent ecological assessments of the continent is reviewed and its compatibility with current models for explaining hunter–gatherer behaviour in Australia is assessed.

THE ECOLOGY OF AUSTRALIA

Morton et al. (2011) outline a series of propositions concerning the ecology of Australia, modified from those originally stated in Stafford Smith and Morton (1990), and review the literature that supports these propositions, paying particular attention to how Australia compares to other arid regions around the world. The propositions are divided into three groups. The first group outlines propositions concerning the physical environment, the second group deals with plants and animals, and the third group discusses human interactions.

Australia is characterised as having an 'infertile, well sorted landscape' (Stafford Smith & Morton 1990: 261). Compared with other parts of the world, Australian rainfall is highly variable, characterised by long periods with little or no rain separated by short periods of very abrupt and at times very heavy rainfall. There is latitudinal variation in the degree of variability in rainfall, with regions south of 27°S showing rainfall variability similar to that of the North American deserts, the Sahel, the northern Sahara and the Karoo (Morton et al. 2011: 316). North of this latitude, rainfall variability is higher than in any other region of the world.

Soils in Australia are depleted in nitrogen and phosphorous (Morton et al. 2011: 317), largely as a result of the tectonic history of Australia. Australia has little relief when compared with the other continents and what volcanism has occurred has, as a consequence, made only local contributions to soil fertility. Unlike the situation on continents in the northern hemisphere, aeolian transport does not contribute greatly to nutrient levels in Australia. However, while the continent has low average soil fertility, there is considerable local variation in nutrients over a variety of different spatial scales from 10^1 to $10^4 m^2$.

In the desert regions, low relief and low fertility combined with intermittent rainfall mean that banded vegetation patterning is common (e.g. various authors in Tongway et al. 2001). Water run-off occurs between the vegetation bands and infiltrates within the bands. The bands are typically aligned along the contour. The distances between bands reflect rainfall amounts, with distances increasing as average annual rainfall declines (Esteban & Fairen 2006). Dunkerley (2002) describes infiltration rates into mulga groves and concludes that it is a combination of nutrient availability and moisture run-on and run-off that is responsible for distances between bands. Similar results are suggested for bluebush shrubland (Dunkerley 2000). In Australia such localised differences in fertility are the dominant pattern throughout large portions of the arid regions.

A second group of propositions deals with plant and animal life. Differences between Australia and other regions are well illustrated by the nutrient-poverty/intense fire theory (Orians & Milewski 2007). The theory suggests that plants growing in soils that are poor in nutrients, like those in Australia, produce more carbohydrate than expected, and this abundance manifests in the form of hardwood, exudates such as nectar and sap, and other tissues that are hard for herbivores to digest. Materials are abundant because they are not consumed and, therefore, they provide the fuel for intense fires. When fires occur some of the nutrients are volatilised, continuing the cycle of nutrient depletion. The periodic nature of fires also produces alternating abundance and dearth of nutrients in the form of ash, with a periodicity that, like rainfall, is difficult to predict. One of the consequences of the nature of vegetation in Australia is, therefore, that consumption by herbivores and detritivores is at times replaced by destruction by fire.

High carbohydrate production combined with a reduced turnover of the long-life leaves on Australian plants produces a high standing biomass of perennial plants. Figures range from 5000kg/ha for chenopod shrublands through to 55,000kg/ha for the semi-arid *Eucalyptus* woodland (Morton et al. 2011). Grass build-up is pronounced after rainfall, particularly in the north of Australia, and fire is frequent in these areas as a consequence. It is most apparent in *Spinifex* grasslands. Fire affects *Acacia* scrubland but less often, since the build-up of grass biomass is less pronounced.

Herbivores in Australia must deal with poorly digestible and intermittent plant production or become specialised in the use of those perennial plants that are available. As a consequence, groups of animals have developed unusual characteristics. Birds in Australia, for instance, are characterised as opportunistic, prone to irruption and nomadic. Birds are more significant in granivory than populations in North and South America. Morton (1985) compared the relative importance of mammals, ants and birds as grain consumers between the two American continents and Australia and found that, unlike those in North America, mammals in Australia are responsible for negligible consumption of grain. Rodents are also insignificant, although comparisons are impeded by the post-European extinction of many rodent species. Ants consume similar quantities of grain to those in North America, but the overall grain predation in Australia is less than in North America.

Morton proposes that birds—the pigeons, parrots and finches that are ubiquitous and abundant in Australia—may be the highly mobile equivalent of rodents. As will be discussed below, these differences in granivory by Australian versus American species are interesting for archaeologists because seed eating was an important part of some arid-zone Aboriginal subsistence strategies.

The large herbivores—kangaroos and wallabies—are affected by the intermittent nature of rainfall and, therefore, production. McCarthy (1996), for instance, considered the red kangaroo (*Macropus rufus*) in an attempt to determine sustainable culling rates. He was able to show that populations were better modelled if the environment was treated stochastically, that is, was subject to random variation in the availability of resources. He found that there is no simple relationship between rainfall and population size. In South Australia summer rainfall correlates with the availability of food and hence the survival of young and old alike. However, the previous year's rainfall is also implicated and, at a smaller spatial scale, migration to zones with better resources is also important. As the McCarthy study shows, simple deterministic models that seek to predict population numbers from one year to the next are misleading.

Heavy rainfall produces extensive lake systems that can attract huge numbers of birds for extended periods; however, these rainfall events are infrequent, driven since at least the mid-Holocene by variations in the El Niño–Southern Oscillation (ENSO) cycles. Kingsford and Norman (2002) describe how many waterbirds respond to the presence of these habitats, where they feed and breed, and that when the habitats become depleted they disperse or die. In the northern hemisphere abundance of waterbirds is predictable, but in Australia numbers are dependent on available habitat and food resources, both of which are temporally variable. Concentrations of waterbirds at any single location are low immediately after flooding, since they are able to disperse to multiple habitats, but are higher during dry times, when habitat is spatially limited. Movement of waterbirds varies from predictable to nomadic, with inland areas being the least predictable. Dry periods that follow flooding mean birds must move, in some instances covering distances greater than 1000km. Some birds do not move. Kingsford and Norman (2002) describe a study by Barnard (1927), who observed pelicans that remained at one location during a dry period and perished. As they point out, there is an energy conflict between moving to a new location and remaining in one place hoping for new resources to become available. For some Australian waterbirds rains stimulate the breeding cycle, with the corollary that during droughts breeding effectively stops or is very much reduced. Breeding is related to both food availability and habitat, something that is seasonally predictable outside Australia but much less so within the continent.

The final proposition put forward by Morton et al. (2011) concerns people. This is an addition to the propositions in the original study (Stafford

Smith & Morton 1990), where people were not mentioned. Morton and colleagues propose that people enhanced production and stabilised access to resources and that this had an effect on the environment. Population shifts are linked to environmental change, with population expansion and contraction from and to refugia but an overall stepwise increase in population and a marked increase over the last 1000 years. Studies are cited that indicate the environmental effect of people hunting, burning, irrigating and dispersing plant propagules. The use of fire is given prominence, particularly its use in *Spinifex* grasslands. Based on a report by Walsh (1990), it is suggested that sowing and planting was a regular activity and that some plant species were dispersed over hundreds of kilometres.

The archaeological evidence for human–environment interaction is discussed in more detail in the next section. Despite the Walsh study, it would be wrong to assume that Aboriginal people in the arid regions of Australia were manipulating plants in ways that in other parts of the world led to their domestication. Some manipulation of the environment occurred, notably by fire, but the extent of this manipulation both in time and space needs to be considered in relation to arid-zone population levels. When discussing animals, Morton and colleagues emphasise that the lack of continuity and predictability of the Australian climate combines with the geology of the continent to produce aspects of the ecology that, to use their terminology, accentuate the differences with other regions of the world (Morton et al. 2011: table 3). Australia has a unique ecology because the combination of low fertility and intermittent rains produced a flora with an excess of carbohydrate manifested in such things as hardwood and sap. The fauna that evolved reflected these aspects of the environment. The people who settled Australia had to deal with this unique ecology using, as will be argued below, a mixture of strategies, some with superficial similarities to those used by other species. The difference, however, was that these were culturally and technologically aided strategic solutions. This greatly expanded both the type and range of solutions available to people compared with other animals. There is a fascinating story to tell about the intricacies of the solutions that Australian Aboriginal people developed through time to the challenges provided by the changing Australian environment. The key to understanding this story is to understand temporal processes in the archaeological record much in the way that Morton and colleagues emphasise the importance of recognising temporality in the environmental processes present in Australia.

ETHNOGRAPHIC OBSERVATIONS

Ethnographic studies attest to the types of plant resources that were exploited by Aboriginal people, with a number of studies providing extensive lists of species that were used (e.g. Gould 1969; O'Connell et al. 1983;

Cane 1987; Walsh 1987; Veth & Walsh 1988). O'Connell et al. (1983), for instance, list 92 plant species used by the Alyawara, 36 used for seeds, 8 for roots, 26 for fruit and 32 for a variety of other products, including nectars, resins and flowers. The concentration on seeds, fruits and nectars rather than tubers reflects the ecology of plants in the arid areas of Australia.

The best calorific returns come from fruits and tubers, with grasses producing lower net returns because of high processing costs (O'Connell et al. 1983; Edwards & O'Connell 1995). O'Connell and Hawkes (1984) describe how Alyawara women passed by abundant grass seed resources, preferring higher caloric yields from tubers, fruits and larvae, and how these choices are consistent with optimal foraging strategies. Observations concerning male hunting were analysed using the same approach. These studies greatly influenced archaeological analyses. However, the ethnographic studies were of necessity conducted over relatively short periods of time. The Alyawara observations, for instance, were made over 16 months in 1973–75. This permitted a number of separate foraging and hunting trips to be recorded but was not long enough to encompass the range of environmental changes typical for arid regions. Some sense of the impact of environmental variability through time is provided by the comment by O'Connell and Hawkes that heavy rains in 1974 promoted extensive seed production by several species of *Acacia*. But there was no opportunity to view such shifts in production over a number of years, nor were declines associated with prolonged droughts investigated. Instead, O'Connell and Hawkes described hunting and foraging forays to a variety of habitat 'patches.' Archaeologists used the optimal foraging approach introduced by O'Connell and Hawkes to understand the nature of sites in different locations (discussed below). However, in doing so they continued the short-term nature of explanation introduced by the ethnographic studies.

Contemporary observations of resource gathering by Aboriginal people are complicated by the availability of store-bought foods that in many communities reduced but did not eliminate the use of native resources (Meehan 1989). The availability of vehicles and rifles had an impact on hunting and gathering returns by extending the range of locations that might be accessed within a particular time frame (O'Connell & Hawkes 1981). There are different strategies for dealing with opportunistic resource abundance in both harvest size and chances of failure. Bliege-Bird and Bird (2008) use honey as an example of an opportunistic resource with a high variance in reward. There is some chance that the gatherer will return with a comb that is several times the typical average size. Goanna is an example of a low-variance target. In the Western Desert foragers will normally return with four to five individuals, but they only very exceptionally return with either 10 individuals or with nothing. For a variance-prone forager, the benefits of a large harvest outweigh the risks of returning empty-handed. For a risk-averse forager, returning empty-handed is of more significance than the possible benefit from gaining a large foraging return. Twentieth-century observations of Aboriginal foraging practices speak to the range of plants and animals that were used both today and in the past, but where vehicles and store-bought foods are present it is likely that a level of predictability has changed. Purchased food sources have not replaced wild sources in some Aboriginal outstation communities but may serve as a buffer against uncertainty. Purchased staples such as flour and sugar, as well as canned goods, can be relied on during periods of prolonged shortage, whereas wild sources cannot. In this sense purchased food sources benefit risk-taking behaviour since there is always a purchased alternative if foraging expeditions prove fruitless. Vehicles greatly increase the ease with which distant sources can be accessed and decrease time spent in unproductive areas between resources. Both these factors have had an effect on the economy and, therefore, can be expected to have had an effect on the nature of occupation compared with that in pre–European contact periods (Altman 1984; Meehan 1989; Balme 2008). Certainly, modern-day Aboriginal people make economically sensible decisions when foraging and hunting that can be analysed with techniques such as optimal foraging. But if the timescale of observation is extended, then the really interesting changes are those that have changed the nature of hunting and gathering activity. Guns, vehicles and store-bought foods change the level of certainty and, therefore, can be expected to have influenced the whole nature of the socioeconomy.

ARCHAEOLOGICAL STUDIES OF HUNTER–GATHERERS

Binford's studies in the late 1970s and 1980s produced a series of conceptual tools useful for archaeologists interested in hunter–gatherers. Binford studied the Nunamuit from Alaska, who occupied a highly seasonal environment. Resources varied by season and Binford (1980) was able to show how seasonal access to resources was related to both the nature of settlement and to the types of material culture used at a particular location or transported from place to place. He distinguished between 'foragers,' employing residential mobility to move consumers between patches of low-abundance resources, and 'collectors,' whose mobility is tethered to stores and who employ logistical mobility to provision consumers from widely dispersed, seasonally superabundant patches. To aid his discussion, he offered a typology of sites definable by the types of artefacts left at different locations. Because the environment was markedly seasonal, movement between different seasonally available resources structured space. Binford (1981) argued that the archaeological record was, therefore, the outcome of the operation of a cultural system operating over medium- to long-term time periods. However, many archaeologists, in applying Binford's theories, ignored the notion of a cultural system and simply applied Binford's site types directly to interpret the archaeological record. As a consequence the spatial distribution of archaeological sites was emphasised over their temporality

(Holdaway & Wandsnider 2006). In many ways this was similar to earlier approaches, like those of Thomas (1973, 1975) in the Great Basin, that had sought to understand land use across wide geographic areas.

Beginning in the 1990s, Binford's work was integrated with studies based on optimal foraging theory and behavioural ecology (e.g. Torrence 1989; Bettinger 1993). Archaeologists described past hunter–gatherer behaviour using the concept of 'strategies,' similar in form to the analytical evolutionary stable strategies of evolutionary ecology. Mobility strategies (e.g. Amick 1996; Smith & McNees 1999; Bamforth & Becker 2000), technological and land use strategies (e.g. Cowan 1999), reproductive strategies (e.g. Bettinger 1993) and subsistence strategies (e.g. Dering 1999; Stafford et al. 2000) were all recognised. However, this approach further diluted Binford's interest in long-term cultural systems and replaced it with typologies of hunter–gatherer sites and strategies. Studies appeared that categorised archaeological evidence as either forager or collector adaptation (e.g. Sanger 1996; Cowan 1999). Other studies used the site types (residential camps, field camps, stations, caches and locations) that Binford had identified (Simms 1992). The definition of strategies in the studies cited above created high-level categories of hunter–gatherer behaviour apparently applicable to many different places (Holdaway & Wandsnider 2006).

In Australia the distribution of archaeological sites was also related to seasonality through rainfall. During wet periods, populations dispersed to exploit resources in regions where water sources were ephemeral. As the rains departed and drought prevailed, people moved to more permanent water sources and used local resources. This interpretation formed the basis for explanations of the distribution of archaeological sites in many studies (e.g. White & Peterson 1969; Allen 1972; Ross 1984; Williams 1987; Smith 1989, 1993, 1996; Ross et al. 1992; Veth 1993; Barton 2003). The artefact contents of sites were said to reflect the relative distance to water, which in turn affected the nature of activities undertaken. Sites located away from sources of permanent water should show relatively few artefacts with little evidence for maintenance activities, while those closer to more permanent water should have more abundant artefacts and show a greater range of materials and artefacts, reflecting occupation by larger groups for longer periods (e.g. Veth 1993: 71). Thus seasonality in the form of rainfall was related to the spatial distribution of sites and, via their contents, to the identification of site types. However, despite the difference in environment in Australia as outlined above, the typology of site types closely resembled that proposed by Binford (e.g. Barton 2003). Like other hunter–gatherer studies those related to the Australian material were largely atemporal. Time was subsumed by the temporality of seasonal changes.

When time is reduced to seasonal shifts, change becomes hard to demonstrate, with the result that those changes that can be seen are punctuated. Much of this relates to a mismatch in the temporal scale of evidence wherein human behavioural changes are related to landscape changes operating over geological time periods (Allen et al. 2008; Holdaway & Fanning 2010). Hiscock, for instance, using behavioural ecology, has related changes in both technology (e.g. Hiscock 1996) and assemblage composition (e.g. Hiscock 1994) to mid-Holocene movements into arid regions in Australia. In the former study he cites changing strategies as the reason for differences in the degree of bipolar flaking present in sites in the north of Australia. In the later study he explains the presence of a range of new artefact forms (adze bits, backed blades and seed-grinding gear) as adaptations to the risks involved in moving into new, particularly arid environments. In both studies change is related to shifts in near continental patterns of environmental change.

The problems involved in understanding the significance of the changes identified in studies like those of Hiscock are well illustrated by the study of red kangaroos by McCarthy (1996) cited above. Depending on both the spatial and temporal scale of observation, McCarthy was able to show that the correlation between kangaroo numbers and rainfall changed. He further showed that a stochastic model for measuring change was a better way of modelling the dynamic changes through time in kangaroo population numbers. Because archaeologists have relied on a combination of land use models based on seasonal shifts, evolutionary ecological studies based on short-term ethnographic observations, and ethnographically derived functional site types, the random shifts in environment discussed by Morton and colleagues (2011) have been much less emphasised. As a consequence, Aboriginal history is seen to be composed of a small number of dramatic changes, like those discussed by Hiscock (1994, 1996), separated by periods of prolonged stasis. Like the ethnographic studies of modern Aboriginal subsistence that ignore the consequence of historical change, what is missing from archaeological studies is a sense of the implication for Australian ecology of dynamic environmental change that is the key theme in the non-human ecological studies discussed by Morton and colleagues above. Alternative approaches are needed, like those we have applied to the Late Holocene archaeology of western New South Wales.

ALTERNATIVE APPROACHES: THE ARCHAEOLOGY OF WESTERN NEW SOUTH WALES

Western New South Wales has an average annual rainfall of less than 250mm and pan evaporation exceeding 2000mm and, like other parts of Australia, very high coefficients of variability for these parameters. Most Holocene palaeoenvironmental evidence suggests that a climatic optimum occurred around 4000 BP, desiccation and aridity from about 3000 BP to about 1500 BP, followed by amelioration up to the present (Holdaway et al. 2002, 2010a). The majority of available water is found in ephemeral creeks, rock holes and other features that hold water for varying lengths of time following rain. Drought conditions prevail but are broken by the punctuated and highly localised rainfall of the type discussed by Morton and colleagues. Areas receiving rain consequently experience brief increases in water availability and relative, though limited, increases in land productivity, while adjacent areas remain dry (Fanning et al. 2007).

This environmental history can be seen by analysing radiocarbon age determinations obtained from the heat-retainer hearths that are commonly associated with stone artefact concentrations in the region. In the Rutherfords Creek catchment, for example, statistically significant correlations can be demonstrated between the radiocarbon ages and peaks in the record of sea surface temperatures (SSTs) derived from deep-sea cores that provide an indication of eastern Asian monsoon activity, as well as shifts in the position of the Inter-Tropical Convergence Zone (ITCZ) (Holdaway et al. 2010a). Changes in these systems have a direct impact on Australian continental rainfall, where the southward positioning of the ITCZ and periods of higher SSTs across northern Australia are associated with higher rainfall in northern and central Australia (Sturman & Tapper 2006). In contrast, the same hearth radiocarbon ages show negative correlations with the record of dust movement from the Australian continent retrieved from an ombrotrophic peat bog in the Old Man Range, Central Otago, New Zealand (Marx et al. 2009). Hearths were constructed more frequently during periods with higher moisture levels but were constructed less frequently when dust transport was higher and conditions in the interior were likely to have been more arid. A marked variation in occupation intensity is suggested, from little or no occupation during dry periods to more frequent activity during more moist times (Holdaway et al. 2010a). However, the timing and degree of these environmental changes would have made the prediction of resource abundance particularly difficult for the Aboriginal occupants of this landscape. Like the non-human faunal examples discussed above, dealing with this unpredictability was a key issue for Aboriginal groups who occupied the region.

Short-term increases in the availability of water and food resources initiated by unpredictable and localised rainfall produced, over the long term, a discontinuous archaeological record in many parts of western New South Wales. Dense artefact concentrations near watercourses are separated by diffuse artefact scatters and isolated occurrences. The sparse and unpredictable character of resource distributions in the short term meant that it was neither easy to spatially target foraging expeditions nor possible to know precisely where stone suitable for the production of artefacts might be found. While raw material is locally abundant (cobbles of silcrete and quartz are readily available in dry creek channels and as extensive stone pavements on adjacent slopes, and silcrete cobbles also occur in the form of boulder mantled outcroppings; Douglass & Holdaway 2011), stopping to find and produce tools as each foraging opportunity was encountered had a time and energy cost. The response was a technological strategy based on a generalised tool kit (Kuhn 1994) carried as a 'hedge' against unforeseen needs, but one that did not include large quantities of retouched tools.

Retouch is rare on stone artefacts in New South Wales and there is little indication of substantial resharpening. Even the retouch on typologically defined scrapers is not very invasive. The exceptions are tula adze slugs (bits for flake adzes; Holdaway & Stern 2004), but these account for only a small proportion of the total number of tools. However, while retouched tools are rare, flakes are abundant. Most flakes were struck in a single direction, although the dorsal scars on flakes indicate some flaking from opposed platforms and some indication of core rotation (Holdaway et al. 2004). Despite the apparent lack of complexity in technology, assemblages show evidence for the selective removal of larger flake blanks and the transportation of these flakes over a variety of distances.

Movement of flakes can be assessed using a method based on the proportion of cortex present on artefacts (Dibble et al. 2005; Douglass et al. 2008; Holdaway et al. 2010b; Lin et al. 2010; Douglass & Holdaway 2011). Application of the method to assemblages from western New South Wales returns values of the cortex ratio consistently below one, indicating that cortex is under-represented (Douglass et al. 2008). The addition of many non-cortical flakes and cores to assemblages would explain the observed values of the cortex ratio. However, the local abundance of raw materials, as well as surveys of the size of stone cobbles, makes the scenario of extensive raw material importation seem unlikely (Douglass & Holdaway 2011). A more probable explanation for the disparity in cortex proportions is the removal of material from assemblages for use elsewhere. Cortex ratios less than one reflect a tendency towards the removal of artefacts with a greater cortical surface area to volume ratio than the nodules from which they were produced. This would result from the selective removal of large blanks, blanks that would tend to have cortex on their dorsal surface as a consequence of their size (Douglass et al. 2008). Artefacts that have a high proportion of cortex and large surface area, but that are also thin and therefore have a low volume, most affect the cortex ratio. The ratio, therefore, also informs on the shape of the flakes that were removed, that is, those having a high edge to mass ratio.

If, for example, the cortex ratio calculated for the Rutherfords Creek assemblages is plotted against the spatial location of the assemblages, the nature of movement can be determined (Holdaway et al. 2012). Movement frequency (both on an individual and group level) affects the potential for artefacts to be moved, in that less movement will result in more artefacts discarded where they were manufactured and greater movement will increase the probability that any one artefact is moved away from its manufacturing location. Therefore, all things being equal, fewer moves will tend to produce increased cortex ratio values and more moves the opposite. Movement linearity will affect the linear distance that an artefact will be moved. Movements that are not linear may cover a great total distance but not move the artefact far from the place where it was produced. Linearity can be expressed using the concept of tortuosity. The degree of movement tortuosity ranges from a non-tortuous straight-line movement between points to a movement path so tortuous as to cover a plane without crossing itself. Differences in movement tortuosity, therefore, represent differences in the thoroughness of landscape usage (Roshier et al. 2008). Low tortuosity (i.e. more linear movement paths) is associated with a higher velocity of movement across the landscape, leading to lower values for the cortex ratio at any one location. In contrast, high tortuosity, while still potentially involving the transport of flakes, will lead to local deposition and, therefore, an increased chance that an artefact assemblage at any one location will have a higher cortex ratio value. Finally, artefact longevity will have an effect on artefact movement since those artefacts that exist for longer periods of time before being discarded will have a greater chance of being transported over greater distances.

Rutherfords Creek flows into Peery Lake, a large ephemeral lake basin that, when flooded, supports a diversity of resources such as fish, birds and grazing animals. What is striking about the distribution of cortex ratio values calculated for individual assemblages in Rutherfords Creek is that there is no uniform gradation of values towards or away from the lake (Figure 4.1). The majority of the assemblages have cortex values well below one, suggesting considerable artefact movement. Despite the existence of the lake 'resource patch,' the cortex ratio values suggest that tortuosity of movement was relatively low and velocity high.

Both the low cortex ratio values displayed by the artefact assemblages and the results of the hearth radiocarbon ages previously described suggest a series of episodic occupations. People who came to Rutherfords Creek made hearths and manufactured artefacts. But they did this only intermittently and when they left, carrying the most efficient stone artefacts possible, they did not return until environmental conditions were suitable. The large, thin flakes that they transported were not designed for use in the places where the artefacts were manufactured but for use at the 'in-between' places in the wider Australian landscape. The archaeological record is not, therefore, interpretable as the result of a uniform seasonal round with established site types, even if 'seasonality' is related to water availability. As others have shown, there is no simple relationship between the size and complexity of assemblages and increasing stream order (Shiner 2008). Indeed, in the Shiner study, assemblage variability contrasted dramatically with the expected array of spatially distinct

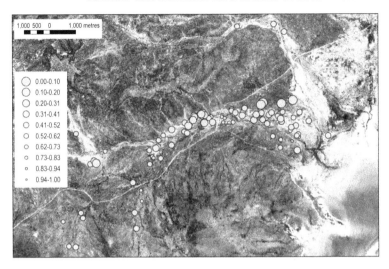

Figure 4.1   An aerial image of the Rutherfords Creek and Howells Creek catchments draining into Peery Lake in western New South Wales. Cortex ratio values for artefact assemblages on randomly selected scalds (bare soil patches with high artefact visibility) are shown by dots

assemblage types. Instead, assemblages from a series of microenvironmental locations showed uniformly high variability, reflecting the relatively long period of time available for the variability to accumulate as a result of stochastic environmental processes like those discussed by Morton and colleagues.

Seen from this perspective, Aboriginal people used mobility in a superficially similar way to other Australian fauna. Faced with a variable and unpredictable resource base, one solution was to move in search of locally abundant resources. Of course, the resemblance to animal behaviour is only superficial because Aboriginal people had at their disposal a complex technology and set of cultural behaviours that the non-human species did not. That provided them with a variety of responses to any one set of conditions. This variability provides the content for the history of the development and success of Aboriginal people in Australia. It is a history that is much more intriguing than the punctuated view put forward by the inadequate, de-temporalised set of archaeological interpretative models currently available.

CONCLUSION

Twenty years on from the original Stafford Smith and Morton (1990) paper, the ecological understanding of what makes Australian ecology different from that of the other continents is more refined (Morton et al. 2011). An infertile land subject to unpredictable and variable rainfall developed a unique flora. This in turn supported a fauna with significant differences from that found in other regions of the world. When they colonised this continent, the ancestors of present Aboriginal people were faced with a series of challenges. That they solved them is self-evident; Australia's vibrant Aboriginal peoples and their rich and diverse culture are testament to their success. However, there is a history to this success that is important to understand. It should not be characterised as a few punctuated changes but as an interaction with a dynamic, unpredictable environment. There is always the temptation to de-temporalise the archaeological record by failing to take account of changes that occur at different times. In the Australian case this temptation must be strongly resisted because dramatic environmental shifts are such a key aspect of the continent's ecology. In archaeology this means that models derived from outside Australia may not always be successfully applied. However, as we have demonstrated in this chapter, there are techniques that can bring the dynamic nature of the past to life.

REFERENCES

Allen, H. 1972. Where the crow flies backwards. Unpublished PhD dissertation, University of Sydney.

Allen, H., S.J. Holdaway, P.C. Fanning & J. Littleton. 2008. Footprints in the sand: appraising the archaeology of the Willandra Lakes, western New South Wales, Australia. *Antiquity* 82: 11–24.

Altman, J.C. 1984. The dietary utilisation of flora and fauna by contemporary hunter-gatherers at Momega outstation, north-central Arnhem Land. *Australian Aboriginal Studies* 1: 35–46.

Amick, D.C. 1996. Regional patterns of Folsom mobility and land use in the American southwest. *World Archaeology* 27: 411–26.

Balme, J. 2008. Comment. *Current Anthropology* 49(4): 679–80.

Bamforth, D.B. & M.S. Becker. 2000. Core/biface ratios, mobility, refitting, and artifact use-lives: a paleoindian example. *Plains Anthropologist* 45: 273–90.

Barnard, H.G. 1927. Effects of drought on bird life in central Queensland. *Emu* 27: 35–7.

Barton, H. 2003. The thin film of human action: interpretations of arid zone archaeology. *Australian Archaeology* 57: 32–41.

Bettinger, R. 1993. Doing Great Basin archaeology recently: coping with variability. *Journal of Archaeological Research* 1: 43–66.

Binford, L. 1979. Organization and formation processes: looking at curated technologies. *Journal of Anthropological Research* 35: 255–73.

Binford, L. 1980. Willow smoke and dogs' tails: hunter-gatherer settlement systems and archaeological site formation. *American Antiquity* 45: 4–20.

Binford, L. 1981. Behavioral archaeology and the "Pompeii Premise." *Journal of Anthropological Research* 37: 195–208.

Bliege-Bird, R. & D.W. Bird. 2008. Why women hunt: risk and contemporary foraging in a Western Desert Aboriginal community. *Current Anthropology* 49(4): 655–93.

Cane, S. 1987. Australian Aboriginal subsistence in the Western Desert. *Human Ecology* 15: 391–434.

Cowan, F.L. 1999. Making sense of flake scatters: lithic technological strategies and mobility. *American Antiquity* 64: 593–607.

Dering, P. 1999. Earth-oven plant processing in Archaic Period economies: an example from a semi-arid savannah in south-central North America. *American Antiquity* 64: 659–74.

Dibble, H.L., U.A. Schurmans, R.P. Iovita & M.V. McLaughlin. 2005. The measurement and interpretation of cortex in lithic assemblages. *American Antiquity* 70(3): 545–60.

Douglass, M.J. & S.J. Holdaway. 2011. Quantifying stone raw material size distributions: investigating cortex proportions in lithic assemblages from western New South Wales. In J. Specht & R. Torrence (ed.), *Changing perspectives in Australian archaeology*: 45–57. Technical Reports of the Australian Museum, Online 23.

Douglass, M.J., S.J. Holdaway, P.C. Fanning & J.I. Shiner. 2008. An assessment and archaeological application of cortex measurement in lithic assemblages. *American Antiquity* 73: 513–26.

Dunkerley, D.L. 2000. Hydrologic effects of dryland shrubs: defining the spatial extent of modified soil water uptake rates at an Australian desert site. *Journal of Arid Environments* 45: 159–72.

Dunkerley, D.L. 2002. Infiltration rates and soil moisture in a groved mulga community near Alice Springs, arid central Australia: evidence for complex internal rainwater redistribution in a runoff–runon landscape. *Journal of Arid Environments* 51(2): 199–219.

Edwards, D.A. & J.F. O'Connell. 1995. Broad spectrum diets in arid Australia. *Antiquity* 69: 769–83.

Esteban, J. & V. Fairen. 2006. Self-organized formation of banded vegetation patterns in semi-arid regions: a model. *Ecological Complexity* 3(2): 109–18.

Fanning, P.C., S.J. Holdaway & E. Rhodes. 2007. A geomorphic framework for understanding the surface archaeological record in arid environments. *Geodinamica Acta* 20(4): 275–86.

Gould, R.A. 1969. Subsistence behaviour among the Western Desert Aborigines of Australia. *Oceania* 39: 253–74.

Gould, R.A. 1991. Arid-land foraging as seen from Australia: adaptive models and behavioral realities. *Oceania* 62(1): 12–33.

Hayden, B. 1979. *Paleolithic reflections: lithic technology and ethnographic excavation among the Australian Aborigines.* New Jersey: Humanities Press.

Hiscock, P. 1994. Technological responses to risk in Holocene Australia. *Journal of World Prehistory* 8: 267–92.

Hiscock, P. 1996. Mobility and technology in the Kakadu coastal wetlands. *Bulletin of the Indo-Pacific Prehistory Association* 15: 151–7.

Holdaway, S.J., M.J. Douglass & P.C. Fanning. 2012. Landscape scale and human mobility: geoarchaeological evidence from Rutherfords Creek, New South Wales, Australia, in S.J. Kluiving & E.B. Guttmann-Bond (ed.) *Landscape archaeology between art and science from a multi- to an interdisciplinary approach*: 279–94. Amsterdam: Amsterdam University Press.

Holdaway, S.J. & P.C. Fanning. 2010. Geoarchaeology in Australia: understanding human–environment interactions, in P. Bishop & B. Pillans (ed.) *Australian landscapes*: 71–85 (Geological Society of London Special Publication 346). London: Geological Society of London.

Holdaway, S.J., P.C. Fanning, E. Rhodes, S. Marx, B. Floyd & M. Douglass. 2010a. Human response to palaeoenvironmental change and the question of temporal scale. *Palaeogeography, Palaeoclimatology, Palaeoecology* 292: 192–200.

Holdaway, S.J., P.C. Fanning, D. Witter, M. Jones, G. Nicholls & J.I. Shiner. 2002. Variability in the chronology of Late Holocene Aboriginal occupation on the arid margin of southeastern Australia. *Journal of Archaeological Science* 29: 351–63.

Holdaway, S.J., J.I. Shiner & P.C. Fanning. 2004. Hunter-gatherers and the archaeology of discard behavior: an analysis of surface stone artifacts from Sturt National Park, western New South Wales, Australia. *Asian Perspectives* 43: 34–72.

Holdaway, S.J. & N. Stern. 2004. *A record in stone: the study of Australia's flaked stone artefacts*. Melbourne and Canberra: Museum Victoria and Aboriginal Studies Press.

Holdaway, S.J. & L. Wandsnider. 2006. Temporal scales and archaeological landscapes from the Eastern Desert of Australia and intermontane North America, in G. Lock & B.L. Molyneaux (ed.) *Confronting scale in archaeology*: 183–202. New York: Springer Science and Business Media.

Holdaway, S.J., W. Wendrich & R. Phillipps. 2010b. Identifying low-level food producers: detecting mobility from lithics. *Antiquity* 84(323): 185–94.

Kingsford, R.T. & F.I. Norman. 2002. Australian waterbirds—products of the continent's ecology. *Emu* 102(1): 47–69.

Kirch, P.V. 2005. Archaeology and global change: the Holocene record. *Annual Review of Environment and Resources* 30: 409–40.

Kuhn, S.L. 1994. A formal approach to the design and assembly of mobile toolkits. *American Antiquity* 59: 426–42.

Lin, S., M.J. Douglass, S.J. Holdaway & B. Floyd. 2010. The application of 3D laser scanning technology to the assessment of ordinal and mechanical cortex quantification in lithic analysis. *Journal of Archaeological Science* 37: 694–702.

Marx, S.K., H.A. McGown & B.S. Kamber. 2009. Long-range dust transport from eastern Australia: a proxy for Holocene aridity and ENSO-type climate variability. *Earth and Planetary Science Letters* 282: 167–77.

McCarthy, M.A. 1996. Red kangaroo (*Macropus rufus*) dynamics: effects of rainfall, density dependence, harvesting and environmental stochasticity. *Journal of Applied Ecology* 33(1): 45–53.

Meehan, B. 1989. Plant use in a contemporary Aboriginal community and prehistoric implications, in W. Beck, A. Clarke & L. Head (ed.) *Plants in Australian archaeology*: 14–30 (Tempus 1). St Lucia: University of Queensland.

Morton, S.R. 1985. Granivory in arid regions: comparison of Australia with North-America and South-America. *Ecology* 66(6): 1859–66.

Morton, S.R., D.M. Stafford Smith, C.R. Dickman, D.L. Dunkerley, M.H. Friedel, R.R.J. McAllister, J.R.W. Reid, D.A. Roshier, M.A. Smith, F.J. Walsh, G.M. Wardle, I.W. Watson & M. Westoby. 2011. A fresh framework for the ecology of arid Australia. *Journal of Arid Environments* 75(4): 313–29.

O'Connell, J.F. & K. Hawkes. 1981. Alyawara plant use and optimal foraging theory, in B. Winterhalder & E.A. Smith (ed.) *Hunter-gatherer foraging strategies: ethnographic and archaeological analyses*: 99–125. Chicago: University of Chicago Press.

O'Connell, J.F. & K. Hawkes. 1984. Food choice and foraging sites among the Alyawara. *Journal of Anthropological Research* 40(4): 504–35.

O'Connell, J.F., P.K. Latz & P. Barnett. 1983. Traditional and modern plant use among the Alyawara of central Australia. *Economic Botany* 37: 80–109.

Orians, G.H. & A.V. Milewski. 2007. Ecology of Australia: the effects of nutrient-poor soils and intense fires. *Biological Reviews* 82(3): 393–423.
Roshier, D.A., V.A.J. Doerr & E.D. Doerr. 2008. Animal movement in dynamic landscapes: interaction between behavioural strategies and resource distributions. *Oecologia* 156: 465–77.
Ross, A. 1984. If there were water: prehistoric settlement patterns in the Victorian Mallee. Unpublished PhD dissertation, Macquarie University.
Ross, A., T. Donnelly & R. Wasson. 1992. The peopling of the arid zone: human environment interactions, in J. Dodson (ed.) *The naïve lands: prehistory and environmental change in Australia and the Southwest Pacific*: 76–114. Melbourne: Longman Cheshire.
Sanger, D. 1996. Testing the models: hunter-gatherer use of space in the Gulf of Maine, USA. *World Archaeology* 27: 512–26.
Shiner, J.I. 2008. *Place as occupational histories: an investigation of the deflated surface archaeological record of Pine Point and Langwell stations, New South Wales, Australia* (British Archaeological Reports International Series 1763). Oxford: Oxbow.
Simms, S. 1992. Ethnoarchaeology: obnoxious spectator, trivial pursuit, or the keys to a time machine?, in L. Wandsnider (ed.) *Quandaries and quests: visions of archaeology's future*: 186–98. Carbondale: Center for Archaeological Investigation.
Smith, C.S. & McNees, L.M. 1999. Facilities and hunter-gatherer long-term land use patterns: an example from southwest Wyoming. *American Antiquity* 64: 117–36.
Smith, M.A. 1989. The case for a resident human population in the central Australian ranges during full glacial aridity. *Archaeology in Oceania* 24: 93–105.
Smith, M.A. 1993. Biogeography, human ecology and prehistory in the sandridge deserts. *Australian Archaeology* 37: 35–50.
Smith, M.A. 1996. Prehistory and human ecology in central Australia: an archaeological perspective, in S.R. Morton & D.J. Mulvaney (ed.) *Exploring central Australia: society, the environment and the 1894 Horn Expedition*: 61–73. Chipping Norton: Surrey Beatty and Sons.
Stafford, C.R., R.L. Richards & C.M. Anslinger. 2000. The bluegrass fauna and changes in Middle Holocene hunter-gatherer foraging in the southern Midwest. *American Antiquity* 65: 317–36.
Stafford Smith, D.M. & S.R. Morton. 1990. A framework for the ecology of arid Australia. *Journal of Arid Environments* 18(3): 255–78.
Sturman, A.P. & N.J. Tapper. 2006. *The weather and climate of Australia and New Zealand*. Oxford: Oxford University Press.
Thomas, D.H. 1973. An empirical test for Steward's model of Great Basin settlement patterns. *American Antiquity* 38: 155–76.
Thomas, D.H. 1975. Nonsite sampling in archaeology: up the creek without a site?, in J.W. Mueller (ed.) *Sampling in archaeology*: 61–81. Tucson: University of Arizona.
Tongway, D., C. Valentin & J. Seghieri. 2001. *Banded vegetation patterning in arid and semi-arid environments: ecological processes and consequences for management*. New York: Springer.
Torrence, R. 1989. Tools as optimal solutions, in R. Torrence (ed.) *Time, energy and stone tools*: 1–6. Cambridge: Cambridge University Press.
Veth, P. 1993. *Islands in the interior: the dynamics of prehistoric adaptations within the arid zone of Australia* (International Monographs in Prehistory, Archaeological Series 3). Ann Arbor.

Veth, P. & F.J. Walsh. 1988. The concept of 'staple' plant foods in the western desert region of Western Australia. *Australian Aboriginal Studies* 2: 19–25.

Walsh, F.J. 1987. Influence of the spatial and temporal distribution of plant food resources on traditional Martujarra subsistence strategies. *Australian Archaeology* 25: 88–101.

Walsh, F.J. 1990. An ecological study of traditional Aboriginal use of "country": Martu in the Great and Little Sandy Deserts, Western Australia. *Proceedings of the Ecological Society of Australia* 16: 23–37.

White, C. & N. Peterson. 1969. Ethnographic interpretations of the prehistory of Western Arnhem Land. *Southwestern Journal of Anthropology* 25: 45–67.

Williams, E. 1987. Complex hunter-gatherers: a view from Australia. *Antiquity* 61: 310–21.

Yellen, J. 1977. *Archaeological approaches to the present—models for reconstructing the past.* New York: Academic Press.

5   Integrating Hunter–Gatherer Sites, Environments, Technology and Art in Western Victoria

David Frankel and Caroline Bird

INTRODUCTION

While ethnographic observations expose the intimate interplay between recent indigenous Australian patterns of land use and physical and conceptual landscapes, these fundamental relationships are harder to perceive archaeologically. Rosenfeld and Smith (2002) have demonstrated patterned relationships between site use and stone artefacts, environmental systems and art. These reflect specific local changes in central Australia indicative of conceptual and ideographic associations, as well as more mundane economic adaptations (see also David et al. 1994; Taçon & Brockwell 1995; Smith & Ross 2008). In this chapter we attempt another such analysis, suggesting the way in which a diverse array of evidence can be integrated into a model of long-term changes in indigenous patterns of mobility and land use in western Victoria. The starting point is the evidence from excavations at several rockshelters in Gariwerd (the Grampians)—a series of sandstone mountain ranges that rise up to 700m above the surrounding plains (Figure 5.1).

Gariwerd is locally well-known for the painted and stencil art found within many rockshelters, especially as this comprises the great bulk of all rock art in Victoria. This prompted much of the archaeological work in the ranges, including excavations carried out by Peter Coutts and the Victoria Archaeological Survey 30 years ago (Coutts & Lorblanchet 1982), the material from which was made available for study by Aboriginal Affairs Victoria in consultation with the local Aboriginal community (Bird & Frankel 1998, 2005). The evidence from five sites (Billimina, Drual, Mugadgadjin, Jananginj Njaui and Manja) forms the basis of this analysis, tracing changes over some 20,000 years in technology, site use and environment.

ARCHAEOLOGICAL EVIDENCE

Sites and Assemblages

The excavated shelters have shallow deposits, with looser, darker, more organically rich deposits overlying more consolidated clays where all organic

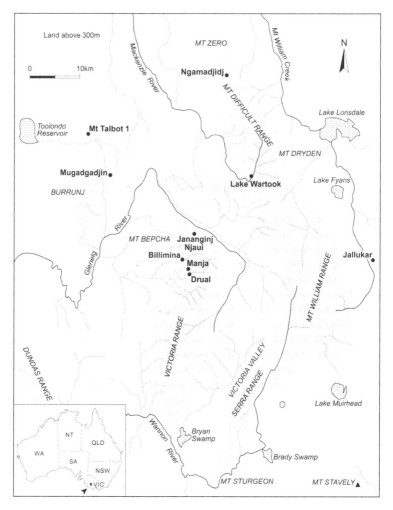

Figure 5.1   Map of Gariwerd showing the location of sites mentioned in the text

material has been leached out and the soil compacted by a variety of processes. Animal disturbance and the generally uneven bedrock exacerbate problems of stratigraphic separation and sequences. This results in a patchy archaeological record, with an uneven representation of periods at different sites and analytical units that vary considerably in integrity and duration (Figure 5.2). Nine assemblages from five sites are available for analysis. They can be grouped chronologically into two main sets, an Early (Late Pleistocene to mid-Holocene) group and a Late (Late Holocene) group. Some differentiation of very recent (Final) material dating to the last millennium is possible at two of the sites.

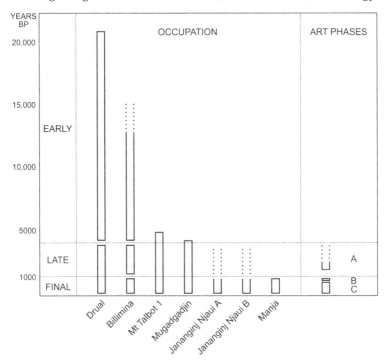

Figure 5.2   Summary of the chronological framework for assemblages and rock art phases

Drual has the earliest archaeological evidence in the region, with basal radiocarbon dates of greater than 20,000 years ago, close to the Last Glacial Maximum. However, no discrimination within the Early material is possible, resulting in the conflation of some 15,000 years of site use. The Late Phase spans the period from about 5000 BP to the abandonment of the shelter, probably in the nineteenth century. Billimina, for which no basal dates could be obtained, was certainly in use by 11,000 years ago and probably for at least several thousand years before that. Here an Early Phase of 6000 years or more can be differentiated from Late (about 5000 BP to 1000 BP) and Final (within the last millennium) Phases. There is less evidence from the other sites, where there is no indication of use before about 5000 years ago.

The rugged terrain and dense vegetation inhibit discovery of open sites (Bird 1989; Edmonds 1995). The only well-documented evidence is from the artificial Lake Wartook, where extensive lithic scatters and hearths (dated to about 4000 years ago) were revealed when the water level was lowered to allow repairs to the dam (Essling 1999; Gunn 2003). This serves, however, to counterbalance the impression that the use of the ranges focused entirely on shelter sites and was always associated with rock art.

Lithic Analysis and Site Use

One striking feature of the Gariwerd assemblages is their diversity in raw materials, technologies and the finished tools. This is evident both in the differences between sites and, in the case of Drual and Billimina, in variation through time.

Raw Materials

The raw materials procured, used and discarded in Gariwerd are diverse in their characteristics and source. Up to a dozen stone raw material types can be distinguished, each with different flaking characteristics. Although precise identification of sources is not possible, these can be grouped to give an indication of major source areas, both within Gariwerd and in surrounding regions:

- The Grampians Group consists of quartzite, sandstone and metasediments producing large flakes or angular chunks.
- The Staverly Complex of cherts, chalcedony, greenstone, non-local quartzite and other fine-grained siliceous rocks outcrop in the Mount Staverly Volcanic Complex to the east of Gariwerd.
- New Volcanics include basalts and dolerite from the south of Gariwerd.
- Quartz occurs both in the ranges and beyond, especially to the north, as veins or as pebbles.
- Southern Murray Basin silcrete is from semi-arid northwest Victoria, north of Gariwerd.
- Rocklands Volcanics are rhyolites, probably from east of Gariwerd.

Assemblage Composition

Although the percentage of quartz in assemblages is variable, all assemblages, with the exception of Manja, contain at least 50% quartz by number. The assemblages broadly fall into two groups: those with about two-thirds (67%) quartz or less, and those with 80% quartz or more. This variation could result from differences in technology or fragmentation. It is clearly not chronological.

The non-quartz components of the assemblages are also diverse. The earlier assemblages tend to be more diverse in their use of stone; the most recent are dominated by local, Grampians Group materials (Figure 5.3).

The variability in raw material composition within the open sites at Lake Wartook may result from different factors. In contrast to the time-averaged

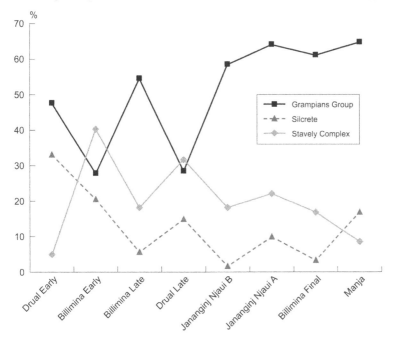

Figure 5.3   Proportions of main raw material types in each assemblage

assemblages from the shelters, these may well represent very specific, discrete or short-term episodes of occupation or tool manufacture.

Formal Tools

Artefacts with evidence of use in the form of secondary retouch or macroscopic edge damage fall into two broad groups: backed tools and edged tools. Backed tools do not occur in the Early Phase. These can be further divided into bondi (or asymmetric) points and geometric microliths. Local quartzite from the Grampians Group is clearly preferred for the former and quartz, volcanic glass or silcrete for the latter. Edged tools include miscellaneous retouched and utilised pieces (about 50% of the total edged tools), pieces with one or more worked notches (about 9%) and scrapers. All raw materials have high proportions of informal edged tools in all periods. The scrapers include both larger examples with convex or straight retouched edges and small convex, 'thumbnail' scrapers, which constitute the largest category of formal tools. These are only made on quartz, fine-grained siliceous materials and volcanic glass and are remarkably uniform through time, from the Pleistocene through to the Final Phase. Other scrapers are more varied in size, and in edge shape and position. The smallest of them overlap with thumbnail scrapers in dimensions but lack their uniformity.

This variation is probably related to raw material characteristics and to resharpening.

Technology

The Gariwerd assemblages can generally be described in terms of a single general reduction sequence, with most technological variation related to raw materials. The clearest change in technology through time is the appearance in the mid-Holocene of specialised burin cores made on large flakes associated with the production of blades for making bondi points.

Site Usage

Tables 5.1a and 5.1b summarise key attributes for each of the lithic assemblages. These show significant variability, although their varied integrity and duration militates against simple direct comparisons. Proxy measures of site usage, such as artefact discard rate, can be employed, although they are affected by raw material and technological variability, with the high fragmentation of quartz and the use of bipolar techniques being the most obvious distorting factors.

Table 5.1a shows a noticeable difference in discard rate between the Early Phase assemblages from Drual and Billimina. This could be attributed to the higher percentage of quartz fragments at Billimina. However, the far lower mean weight of fragments shows that the pieces were more highly reduced, so that the difference is more likely to be one of technology than intensity of site use. This in turn may be related to raw material availability—the quartz used at Drual in the Early Phase is distinctive and probably comes from a particular source. Billimina also differs from Drual in the high proportion of materials (primarily chert) from east of Gariwerd.

Table 5.1a Summary data on stone assemblages: Early assemblages

	Drual	Billimina
Date (range cal BP)	22,000–c. 4000	>10,000–4000
Artefact discard rate per 1000 years corrected for area excavated	7	54 (9945–6665 cal BP) 29 (6665–3980 cal BP)
% quartz	67	85
Mean weight quartz artefacts (g)	1.14	0.35
Total % local materials	83	90
Non-local materials as percentage of non-quartz materials		
Silcrete	33	21
Stavely complex	5	40
Rhyolite/new volcanics	14	7

Table 5.1b Summary data on stone assemblages: Late and Final assemblages

	Mugadg-adjin	Billimina Late	Drual Late	JN A	JN B	Billimina Final	Manja
Date (range cal BP)	?4000–0	?3980–950	?4000–0	?2000–0	2000–0	950–0	<1000–0
Artefact discard rate per 1000 years corrected for area excavated	61	8–35	25	555?	350?	49–158	>2000
% quartz	89	61	81	66	84	66	22
Mean weight quartz artefacts (g)	0.75	0.38	0.38	0.50	0.37	0.51	0.55
Total % local materials	92	90	87	89	95	92	77
Silcrete	62	6	15	10	2	3	17
stavely complex	13	18	32	22	18	17	9
rhyolite/ new volcanics	4	3	22	1	10	5	4
Ratio backed to edged tools	1:1.4	1:0.5	1:4	1:0.5	1:4.3	1:1.3	1:1.1

Artefact discard rates in the later periods also vary (Table 5.1b). Mugadg-adjin and a nearby site at Mount Talbot 1 have higher discard rates than the Late Phase at Billimina and Drual. The Final Phase, however, is character-ised by significantly higher discard rates at both Billimina and Manja. The figures for the two shelters at Jananginj Njaui must be treated with caution, given the uncertainty of dating and duration. However, both assemblages are probably relatively recent, and the discard rate is high. As the rate for the Late and Final assemblages shows no relationship to the incidence or degree of reduction of quartz, it seems likely that changes in it may signal different scales of site usage. Billimina is the only site for which discard rates can be estimated for more than two phases. Here a decline from sometime after 7000 BP is followed by an increase in the last millennium.

Stone artefact manufacture was clearly a significant activity at all sites. The composition of assemblages provides some insights into site function. The ratios of backed to edged tools vary from one backed to more than four edged tools at Jananginj Njaui B to two backed for each edged tool at Billimina in the Late Phase and at Jananginj Njaui A (Table 5.1b). Different modes of explanation for this may be employed which operate at differ-ent scales of time and space, as can be seen in the factors affecting changes at different points in long stratigraphic sequences such as that at Devon Downs in South Australia. Here some changes can be ascribed to local adjustments to resource availability, while others reflect regional patterns of seasonal mobility or even broader, more general technological developments in stone tool types (Frankel 1991: 65–6).

At Graman, in northern New South Wales, McBryde (1977) noted a coincidence of backed pieces and faunal remains. Here the proportions of backed pieces (taken to be components of spears) correlated with those of macropods, suggesting a change in activity related to the availability of different animal species, which in turn was related to changes in the local environment. The shift from the Late to Final Phase at Billimina may signal a similar change in resources and site function in an ecological context. The difference between the two adjacent shelters at Jananginj Njaui may have a different explanation. Where higher levels of maintenance tools are more likely to indicate generalised domestic use, higher proportions of extractive tools such as backed pieces (i.e. spear barbs) suggest a more specialised use. This raises the possibility that backed pieces are markers of men's sites, while more generalised tools indicate more varied occupation. These two modes of explanation are, however, not mutually exclusive. The change at Billimina, therefore, may not be simply an economic one, but may also be related to the broader site function, moving from a special purpose (possibly a men's) site in the Late Phase to a more generalised, multipurpose site in the Final Phase.

ROCK ART

The Gariwerd region contains the densest concentration of rock art in Victoria. More than 100 sites have been recorded (Gunn 1981). These include some larger shelters with hundreds of individual motifs, the most extensive range of which is at Billimina. Most of the sites, however, are small, with a limited number of motifs. The art comprises mostly paintings, although there are also drawings, handprints and hand stencils.

Stylistic Sequence

According to Gunn (1987a, 1987b) superimposition suggests a relative chronology with three distinct phases: the earliest red paintings (Art Phase A), an intermediate phase with red and black drawings (Art Phase B) and the most recent white paintings (Art Phase C). Further support for this sequence comes from the better preservation of white paintings, which commonly have pigment still present, where red paintings are reduced to staining of the rock. Although the radiocarbon dates originally used by Gunn to suggest an absolute chronology for the art are no longer reliable, other evidence reinforces this sequence and suggests a timescale. Surface coatings are common in sites. When associated with art, they invariably cover red paintings and underlie white paintings. These coatings are probably related to an increase in water flow and may be attributed to the increased precipitation of the Late Holocene wetter phase from about 2000 years ago (Gunn 1987a: 66). Art Phase A may, therefore, be regarded as older than 2000 years. In addition, finds of nineteenth-century organic artefacts apparently cached in some shelters with Phase C art tend to confirm the recent date of this art (Gunn 1987a: 45).

Art Phase A: Red Paintings

Gunn suggests a chronological range of about 3500–1600 BP for Phase A art (1987a: 71). It is typified by paintings in red ochre. Bars are the dominant motif, while human figures, lines and emu tracks are also common. Red hand stencils and handprints are also found. Phase A sites occur in local clusters comprising a major site surrounded by several smaller sites. Most sites have one particular dominant motif. With the exception of the ubiquitous bars, motif types also seem to cluster in different parts of the ranges. As most motifs of this phase can be constructed from combinations or variations of bars and emu track motifs, Gunn (1987a: 57; 1987b) has argued that this art forms a 'tightly interconnected system' and that the distinctive distribution patterns indicate that it was closely linked to particular places. The symbols in the art would, therefore, directly relate to mythology tied to named places.

Art Phase B: Red and Black Drawings

Gunn suggests that this art dates from about 1000 years ago (1987a: 71). Art Phase B comprises fine line drawings in red ochre and charcoal. There is a similar range and proportion of motif types to the red paintings of Phase A, with the exception of emu tracks, which are rare. There is, however, less structural similarity between motifs, while the human figures are more animated and less stylised. There is little specialisation in distribution, although there is a concentration of sites with drawings at the north end of the range. Gunn (1987b: 57) speculates that the freer drawing style and looser structure indicate that similar art was produced using other media, such as bark. Alternatively, he suggests that the art was more introspective.

Art Phase C: White Paintings

Gunn suggests that this art dates from about 800 years ago (1987a: 72) and may have continued into the nineteenth century, probably into the contact period. The range of motif types is similar to that used in the other phases, although bars are rare. The motifs are not structurally linked as they were in the red paintings of Phase A, and the technical execution is generally comparatively crude. Art Phase C sites tend to occur singly rather than in clusters. They are also located around the periphery of the ranges in easily accessible locations. This contrasts with the earlier art, which is more widely distributed and often less accessible. There is no evidence of site-specific specialisation of motifs and the range of motifs is evenly distributed.

Understanding the Art

Gunn's analysis (1987a, 1987b) and characterisation of the phases allows their interpretation as distinct periods of artistic activity, while indicating strong continuity between them. He has persuasively argued that the sequence could be related to differences in the social context within which the art was produced. In this way the art of Phase A can be regarded as tightly controlled and associated with particular locations, and was perhaps carried out in specific ritual contexts, while the art of Phases B and C suggests a more casual, public art.

ENVIRONMENTAL HISTORY

The rugged and broken topography of the ranges is characterised by east-facing steep escarpments and shallow slopes to the west. Between these ranges are long, narrow valleys drained by rivers to the north and south, with swampy flats where drainage is restricted. Today Gariwerd has a significantly different climate from that of the surrounding plains, with a relatively high annual rainfall. The complex terrain leads to varied micro-climates. This is reflected in the diversity and richness of its vegetative communities and corresponding resources.

The general pattern of environmental change in south-western Victoria through the Late Pleistocene and Holocene is known from studies of lake levels and microfossils (Jones et al. 1998, 2001) (Table 5.2). At the Last Glacial Maximum, about 18,000 years ago, Gariwerd would have been on the southern edge of an expanded arid zone (Dodson et al. 1992). Small pockets of woodland vegetation may have survived in sheltered places, but the ranges were probably mostly treeless and carried heath or scrub. The valleys would probably have been open grassland with scattered trees. Open grasslands and heath would have persisted throughout the Late Pleistocene, although the diverse topography may well have made parts of Gariwerd a refuge area for tree species. Gariwerd could thus have provided an open and diverse environment with available water, so that the ranges were more attractive than the arid surrounding plains.

As elsewhere in western Victoria, an increase in tree cover followed post-Pleistocene climatic amelioration. The Early Holocene expansion of tree cover in the ranges would have been quite rapid and followed a characteristic succession of *Casuarina*-dominated woodlands followed by eucalypt woodland and grasses. The open red gum woodland with grassy understorey that characterised parts of the Victoria Valley at European settlement may have become established about 7000 years ago. Today the ranges are

Table 5.2 Summary of climatic changes over the last 16,000 years in southwest Victoria, based on modelling of the precipitation/evaporation ratio (after Jones et al. 1998)

Late Pleistocene	16,000–10,500 BP	Dry climate; increasing temperature and precipitation following the Last Glacial Maximum
Early Holocene	10,500–7500 BP	Wetter climate; lake levels began rising about 8500 BP; lakes overflowing by 7000 BP
Mid-Holocene	7000–5500 BP	Wettest period of the Holocene
	5500–3500 BP	Drying conditions
Late Holocene	3100–2000 BP	Dry but unstable conditions
	2000–110 BP	Wetter period
	AD 1840–present	Drier climate, falling lake levels; comparable to Early Holocene

very densely vegetated and relatively inaccessible. However, this probably results from the cessation of Aboriginal burning ('tracks formed by the natives' were identified by early European visitors; C.B. Hall in Bride 1983: 267) and for much of the Holocene the vegetation was probably kept more open than today, particularly in the valleys. A possible exception was during the mid-Holocene wet period, 7000 to 5500 years ago, when the ranges could well have been densely vegetated and relatively inaccessible.

INTEGRATION

Integrating different lines of evidence, from sites, artefacts, art and environment, allows us to identify behavioural changes in time and space.

Billimina provides the best single sequence, with all three art phases represented. The vast majority of the motifs belong to Art Phase A and are, therefore, probably associated with the Late Phase of occupation. This was a period with a relatively low discard rate and a high incidence of backed pieces. It was followed by a period during the last millennium with little art and a lower incidence of backed pieces. That is, there is a systematic variation in the nature of art and the use of the shelter.

At Manja art and archaeological deposition seem to be mutually exclusive. Here nearly all the art belongs to Art Phase A, while the archaeological evidence of occupation falls entirely within the last millennium—that is, painting did not take place at times when the site was in normal domestic use. A similar pattern may explain the predominance of red paintings at both Jananginj Njaui shelters, which have no unequivocal evidence of contemporary occupation.

The evidence from Billimina and Manja fits well with Gunn's interpretation of Art Phase A red paintings as tightly controlled and associated with particular locations, and perhaps carried out in specific ritual contexts. At a time equivalent to Art Phase A these particular sites seem to have been used on a limited basis by specific groups of men (as argued from the dominance of extractive tools). Little or no art production took place during the Final Phase, when the sites had a more domestic character. Only at Drual and Mugadgadjin are there significant numbers of recent white paintings. The range of stone artefacts at these sites can be interpreted as indicating a wider array of activities and more broadly based site use. Once again this fits with Gunn's suggestion that the white paintings of Art Phase C represent a more casual, public art.

It is also possible to relate this sequence to the broad environmental changes from the Last Glacial Maximum through to the recent period. Twenty thousand years ago Gariwerd would have stood out from a semi-arid to arid sand-plain, much as the ranges of central Australia do today. With climatic amelioration towards the end of the Pleistocene the ranges became more heavily vegetated and the plains less arid. At later times, fluctuating levels of precipitation increased water availability both in the ranges and in the rivers, creeks and lakes of the surrounding plains. From a human point of view the place of Gariwerd in the perceived and useable environment must have changed significantly in step with these fluctuations. This can be monitored through consideration of the use of raw materials.

Billimina provides the best basis for exploring change through time. In the Early (i.e. Terminal Pleistocene and Early Holocene) deposits, raw material came from both east and north of the ranges. From about 4000 years ago these materials declined and there was greater focus on local Grampians Group stone. A similar domination of Late and, especially, Final assemblages by local stone is seen at other sites. At Mugadgadjin there is a slightly different pattern, perhaps related to the location of this site in the Burrunj outlier on the western fringe of Gariwerd. Here, silcrete from much farther north of the ranges is dominant in the Late Phase, signalling that Mugadgadjin may have been part of a different land use system, linked more to the adjacent semi-arid regions of the Wimmera and Mallee to the northeast. This perhaps indicates a social pattern not dissimilar to those seen in the nineteenth century when the ranges were at the boundaries of several Aboriginal language groups (Clark 1990), each connected to clans further afield through intermarriage, language and shared matrilineal moiety organisation.

These patterns thus suggest continuously fluctuating strategies of land use and regional associations (Figure 5.4). During the harsher conditions of the Pleistocene, Gariwerd may have functioned as the core territory for groups making less intensive and more wide-ranging use of the arid surrounding plains. It is possible that grinding material found in the Early Phase at Drual may be linked to seed grinding during this period, when grasses were more abundant and other resources scarcer. Forays away from the more reliable

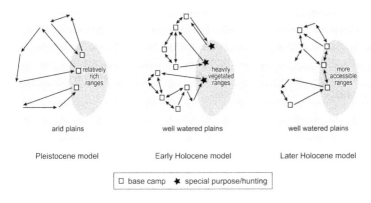

Figure 5.4 Schematic models of land use patterns in and beside the ranges

resources of the ranges are marked, archaeologically, by the acquisition of raw materials from the north and east. With post-Pleistocene climatic amelioration providing more reliable water supplies in the semi-arid regions to the north and allowing the development of wetlands in the surrounding plains, the ranges no longer served as the focus of land use in the same way. It would now have been possible for people to remain out on the surrounding plains, resulting in more independent regional groups. Gariwerd generally can be seen to have become more isolated and more inward looking. One might suggest that this led to the development of somewhat more exclusive social networks, such as those Pardoe (1990, 1995) has suggested for the Murray River.

Some changes in the mid-Holocene may be related to the increased environmental stress brought about by a drier episode from about 5500 to 2000 years ago. This is seen in the fall in discard rates at Billimina. The red paintings of Art Phase A may also be related to this period and associated with a more esoteric use of numerous shelters in all parts of the ranges, which would have been less heavily vegetated and so more easily accessed. Ordinary use of the ranges continued in open country and valleys, and beside swamps and watercourses, as evidenced by a 4000-year-old hearth at Lake Wartook. The use of the ranges changed again in the last millennium. Many smaller, remote shelters, perhaps now more difficult to reach in a more densely vegetated environment, were no longer used at all, while larger ones, especially on the periphery of the ranges, became the focus of general domestic, rather than special, activities. For this period the relative insularity of Gariwerd is signalled by the greater use of local raw materials, and a decline in the use of silcrete and other fine-grained material from beyond the ranges in spite of their better flaking qualities for producing regularly designed tool types.

For Gariwerd, then, a varied array of data on raw material procurement, tool types, site use, art and the environment can be integrated to create a plausible model of fluctuations in the use of the ranges within the natural and culturally perceived environment. The resulting model is one in which responses to climatic circumstances were mediated by historically contingent factors involving pre-existing social relationships and technological processes. This can be seen as leading to periodic reorganisation in the way the ranges and the surrounding plains were integrated within technological systems and regional land use patterns.

REFERENCES

Bird, C.F.M. 1989. Archaeological field survey in the Grampians National Park, 1988. Unpublished report to Parks Victoria, Melbourne.
Bird, C.F.M. & D. Frankel. 1998. University, community and government: developing a collaborative archaeological research project in western Victoria. *Australian Aboriginal Studies* 1998(1): 35–9.
Bird, C.F.M. & D. Frankel. 2005. *An archaeology of Gariwerd. From Pleistocene to Holocene in western Victoria* (Tempus 8). St Lucia: Anthropology Museum, The University of Queensland.
Bride, T.F. (ed.) 1983. *Letters from Victorian pioneers.* South Yarra: Currey O'Niel.
Clark, I.D. 1990. *Aboriginal languages and clans: an historical atlas of western and central Victoria, 1800–1900* (Monash Publications in Geography No. 37). Melbourne: Monash University.
Coutts, P.J.F. & M. Lorblanchet. 1982. *Aboriginals and rock art in the Grampians, Victoria, Australia* (Records of the Victorian Archaeological Survey 12). Melbourne: Victoria Archaeological Survey.
David, B., I.J. McNiven, V. Attenbrow, J. Flood & J. Collins. 1994. Of Lightning Brothers and white cockatoos: dating the antiquity of signifying systems in the Northern Territory, Australia. *Antiquity* 68: 241–51.
Dodson, J., R. Fullagar & L. Head. 1992. Dynamics of environment and people in the forested crescents of temperate Australia, in J. Dodson (ed.) *The naive lands: prehistory and environmental change in Australia and the south-west Pacific*: 115–59. Melbourne: Longman Cheshire.
Edmonds, V. 1995. An archaeological survey of the Serra Range, Grampians (Gariwerd) National Park, south-west Victoria. Unpublished report to Aboriginal Affairs Victoria, Melbourne.
Essling, J. 1999. Analysis of the lithic assemblage from Lake Wartook, Gariwerd, Victoria. Unpublished BA (Honours) thesis, Department of Archaeology, La Trobe University.
Frankel, D. 1991. *Remains to be seen: archaeological insights into Australian prehistory.* Melbourne: Longman Cheshire.
Gunn, R.G. 1981. *The prehistoric rock art sites of Victoria: a catalogue* (Victoria Archaeological Survey Occasional Reports Series 5). Melbourne: Ministry for Conservation, Victoria.
Gunn, R.G. 1987a. Aboriginal rock art of Victoria. Unpublished report to the Victoria Archaeological Survey, Melbourne.
Gunn, R.G. 1987b. The Aboriginal rock art in the Grampians, in *Australia Felix: the Chap Wurrung and Major Mitchell*: 52–63. Dunkeld: Dunkeld and District Historical Museum.
Gunn, R.G. 2003. Three more pieces to the puzzle: Aboriginal occupation of Gariwerd (Grampians) western Victoria. *The Artefact* 26: 32–50.
Jones, R.N., J.M. Bowler & T. McMahon. 1998. A high resolution Holocene record of P/E ratio from closed lakes, western Victoria. *Palaeoclimates* 3: 51–82.

Jones, R.N., T.A. McMahon & J.M. Bowler. 2001. Modelling historical lake levels and recent climatic change at three closed lakes, western Victoria, Australia (c. 1840–1990). *Journal of Hydrology* 246: 159–80.

McBryde, I. 1977. Determinants of assemblage variation in New England prehistory, in R.V.S. Wright (ed.) *Stone tools as cultural markers*: 225–50. Canberra: Australian Institute of Aboriginal Studies.

Pardoe, C. 1990. The demographic basis of human evolution in southeastern Australia, in B. Meehan & N.G. White (ed.) *Hunter-gatherer demography: past and present*: 59–70 (Oceania Monograph 39). Sydney: Oceania Publications.

Pardoe, C. 1995. Riverine, biological and cultural evolution in southeastern Australia. *Antiquity* 69: 696–713.

Rosenfeld, A. & M.A. Smith. 2002. Rock art and the history of Puritjarra rockshelter, Cleland Hills, Australia. *Proceedings of the Prehistoric Society* 68: 103–24.

Smith, M.A. & J. Ross. 2008. What happened at 1500–1000 cal. BP in Central Australia? Timing, impact and archaeological signatures. *The Holocene* 18: 379–88.

Taçon, P.S.C. & S. Brockwell. 1995. Arnhem Land prehistory in landscape, stone and paint. *Antiquity* 69: 676–95.

6 Pushing the Boundaries

Imperial Responses to Environmental Constraints in Early Islamic Afghanistan

David C. Thomas

INTRODUCTION

The prospect of the world's population passing seven billion has refocused attention on the issues of food supply, finite resources and sustainable population levels (see, for example, Brown 2005: 22–39). This chapter explores similar issues in the context of the so-called 'Ghūrid interlude' in Afghanistan (1148–1215 CE; Bosworth 1965: 1103). This period was characterised by the rapid, short-lived expansion of several empires, defined as "any large sovereign political entity whose components are not sovereign" (Taagepera 1978: 113), involving large-scale population movements and associated socioeconomic upheavals. One example of this is the rapid expansion of the Ghūrids (a term used loosely here to include the pre-eminent Shansabānīd dynasty and their rivals), who emerged as a major military power campaigning as far west as Nīshāpūr in Iran and as far east as Bengal. The nature of Ghūrid society, its interrelationship with the environment, and how that society and environment were transformed by their imperial experiment provide the focus of this chapter. Multidisciplinary research in the tradition of the *Annales* School, drawing on archaeological fieldwork, the historical sources and human geography, is ideally suited to investigating the initial success and ultimate failure of the Ghūrid polity to establish a lasting empire.

The limited environmental carrying capacity of the Ghūrids heartland meant that imperial success required access to external resources, both material and human. Like their predecessors, the Ghūrids used the loot and tribute from their raids into the northern Indian subcontinent in particular to expand and aggrandise their urban centres and transform their lifestyle. Recent archaeological fieldwork and the study of satellite images demonstrate that the Ghūrids' summer capital at Fīrūzkūh (modern Djām) is a much larger and more multifaceted archaeological site than has previously been appreciated. Its sudden abandonment following Mongol sieges in 1222–23, and the failure to re-establish a major urban centre at the site, suggests that the Ghūrids were living beyond their means, and that bereft of their external sources of income, they reverted to their traditional, transitory lifestyle.

After outlining the relevant theoretical debates, and the environmental and historical contexts of the study, I will draw on historical and archaeological data to discuss the transformations that the G͟hūrids' lucrative military forays enabled at their summer capital, Fīrūzkūh. The chapter will conclude by assessing how the disintegration of the G͟hūrids' realm affected their heartland.

THEORETICAL DEBATES

Environmental considerations have played a prominent role in the multifaceted philosophical approaches of the *Annales* School over the past 80 years. Early proponents of the *Annales* School drew heavily upon the work of the French human geographer Paul Vidal de la Blache, whose concept of *genres de vie* ('ways of life' or 'lifestyles')

> . . . conveys the notion of an adaptation of the physical *milieu* which is neither totally determined nor totally determinant, but gradually elaborated over the centuries in the light of the body of beliefs and values peculiar to its adherents' *civilisation*. (Lewthwaite 1988: 167)

The most famous *Annales* historian, Fernand Braudel, adopted a more dogmatic approach, arguing that too much emphasis is placed on ultimately insignificant *événements* (short-term, historical and political events) compared with *conjunctures* (medium-term, socioeconomic factors) and *structures* (long-term, slowly changing factors such as geomorphology, environment and climate). Ultimately, in Braudel's opinion (1972: 16), it is the *structures* operating over *la longue durée* that significantly influence the course of history. In response to criticisms that Braudel's approach was environmentally deterministic (Horden & Purcell 2000: 36, 41; Sanderson 1988: 277–8), latter-day *Annales* historians have tended to revert to Vidal de la Blache's more subtle approach to elucidate the complex relationship between environment, economy and social structure (Le Roy Ladurie 1978).

Another criticism of the *Annales* School, and in particular Braudel's work, is its lack of theory relating to social change. One of the School's founders, Lucien Febvre, adopted a Marxist stance; more recently, Emmanuel Le Roy Ladurie (1974) has revisited Malthus's theories on population pressure as a negation of change. Advocates of World Systems Theory (Wallerstein 1976), which is in part inspired by the *Annales* School, have developed more original interpretative models. As a 'world-economy' develops, the socioeconomic gaps between its various components tend to expand. This process promotes instability, but may be masked for a while as technological innovation or the capture of 'new energy supplies' staves off entropic fragmentation (Bintliff 2006: 190). The loot and tribute generated by the

Ghūrid campaigns in the northern Indian subcontinent provide an example of these 'new energy supplies.'

One of the merits of World Systems Theory is that it 'reinvigorates a geographical understanding of politics by according central analytical position to relationships established in space between economically intertwined places' (Smith 2003: 100). Another virtue is the incorporation of variable analytical timescales, which often reveal apparently new features 'to be reincarnations of older structural features or cyclical processes' (Chase-Dunn & Hall 1993: 852). Nomad migrations from the steppe and attempts to plunder the northern Indian subcontinent provide examples of such recurring processes. Despite their relatively short timescales, these *événements* frequently had lasting socioeconomic and environmental impacts, on both the sedentary and nomadic societies—the latter including a multitude of flexible mobile and semimobile subsistence strategies and lifestyles (Cribb 2004: 15–22; see also Mohammed 1973).

MODERN AND HISTORICAL INSIGHTS INTO THE EARLY ISLAMIC ENVIRONMENT

The mountains, river valleys and desert fringes of central and southern Afghanistan formed the core of the Ghūrid empire (Figure 6.1). A significant proportion of the region's early Islamic population, including the Ghūrids' royal court, moved seasonally across these merging environmental zones, utilising the range of resources they offered (Ball 2002: 42–5).

The early Islamic sources reveal little about the Ghūrid heartland itself (Le Strange 1976: 416). It was a remote region, dotted with virtually impregnable fortresses belonging to semi-independent chieftains (al-Yaꜥqūbī, cited in Wheatley 2001: 79–80; Bosworth 1961: 118–19). The rugged terrain provided the Ghūrids with a secure defensive position from which to raid their neighbours (and, periodically, each other). The region's iron ore mines and elevated grazing lands and water sources (Taaffe 1990: 27) also enhanced their military capabilities, enabling them to produce highly prized armour, coats of mail, weapons and horses (Anon 1970: 108; Le Strange 1976: 416–7).

Although the Ghūrid heartland was devoid of large urban centres (Anon 1970: 110; Bosworth 1961: 118), the size and nature of their summer capital Fīrūzkūh (modern Djām) has not been fully appreciated until recently. The location of the winter 'capital' of Zamīn-Dāwar, which was 40 *farsakh* (leagues) from Fīrūzkūh (Juzjani 1970 I: 386), remains unknown, in part because it appears to have been a district rather than a specific place (Kohzad 1957: 32). Two seasons of fieldwork by the Minaret of Jam Archaeological Project at Djām have revealed extensive remains of terraced stone and mud-brick domestic and public architecture beneath the apparently barren, scree-covered mountain slopes around the famous minaret (Thomas 2007)

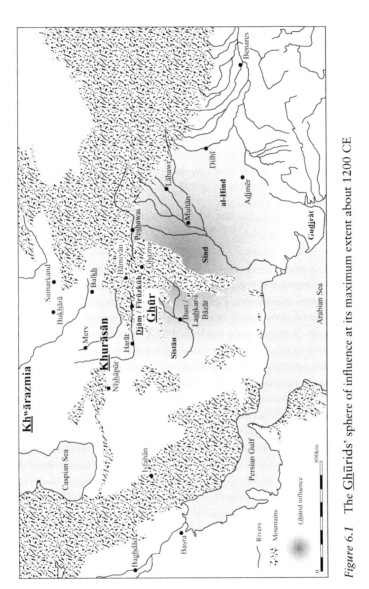

Figure 6.1 The <u>Gh</u>ūrids' sphere of influence at its maximum extent about 1200 CE

Figure 6.2   Minaret of D̲j̲ām and surrounding mountain slopes, pock-marked by robber holes (Photo: David Thomas)

(Figure 6.2). Survey work and detailed analysis of satellite images suggest that the site extends over an area of around 17 hectares, with an estimated peak population of between 1750 and 6900 (Figure 6.3, Table 6.1). Today only a handful of people live in an impoverished hamlet at the site.

The numerous major river valleys that drain the G̲h̲ūrid heartland provided relatively easy access to a ring of urban centres on its periphery and lowlands renowned for their agricultural produce (Le Strange 1976: 410–16). Harāt in the west led to K̲h̲urāsān, while Las̲h̲kar-i Bāzār in the southwest provided access to Sīstān, the 'granary of the East' (Dupree 1977: 43–5; Le Strange 1976: 339). To the south was G̲h̲azna, the gateway to the Indian subcontinent, while Bāmiyān in the east sat on trade routes linking Bal̲k̲h̲ and Transoxania with Kābul and the Indian subcontinent (Figure 6.1). Consequently, although remote, the central location of the G̲h̲ūrid summer capital had numerous advantages in terms of accessibility.

The G̲h̲ūrids were merely the latest of a succession of dynasties (and theologians and traders) to take advantage of the region's pivotal position linking central Asia and the Iranian world with the plains of the northern Indian subcontinent (Ball 2008: 46–7). In the Islamic era, the lands west of the Indus were conquered by the Arab armies around 711–712 CE (Wheatley 2001: 38) and became the economically important province of Sind (Anon 1970: 122). Maritime and overland trade brought perfumes, pearls, diamonds, drugs, textiles and wild animals to the Islamic world, in addition to "religious and political dogmas, artistic ideas, and . . . the human agents who made and traded the objects in which they were manifest" (Flood 2009: 16). Farther east was al-Hind, whose fabulously rich kings and idol

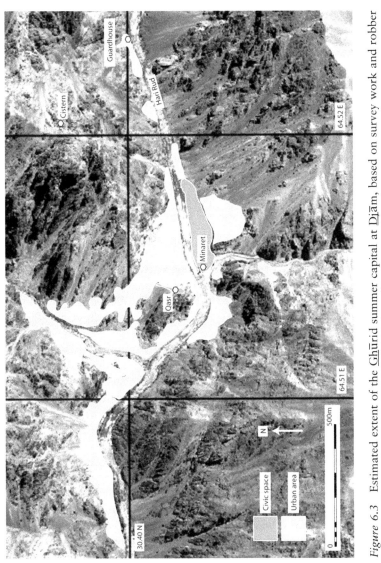

Figure 6.3 Estimated extent of the <u>Gh</u>ūrid summer capital at <u>Dj</u>ām, based on survey work and robber holes identified on the 17 July 2004 Quickbird satellite image

Table 6.1   Population estimates for Djām based on population per hectare, derived from modern ethnographic studies in Iran (Kramer 1982: 125; Sumner 1989: 632–3)

Urban area	17.3ha
100 inhabitants per ha	1750
150 inhabitants per ha	2600
300 inhabitants per ha	5200
400 inhabitants per ha	6900

temples offered the prospect of lucrative returns from military campaigns, such as the 17 raids by MaHmūd of Ghazna in the eleventh century (Wink 1991: 129).

ENVIRONMENTAL CHANGE SINCE THE EARLY ISLAMIC PERIOD

Current debates about climate change and human impact upon the environment highlight the need to synthesise new scientific data with the historical sources and modern observations. Recent climatic studies indicate that a wide range of changes took place in the global climate during the so-called 'Medieval Warm Period' (c. 700–1300 CE; Kremenetski et al. 2004: 114; Trouet et al. 2009: 78). The available data from parts of continental Europe, Fennoscandia and northern Asia point to a two-stage climatic model, with a decrease in summer temperatures after 1100 CE, and increased summer and winter precipitation (Kremenetski et al. 2004: 123). The variability in, and fluctuating nature of, the record, however, illustrates the perils of attempting to correlate regionally and temporally disparate trends with specific historical events over vast landscapes.

The archaeobotanical and -zoological samples collected at Djām provide the only environmental data from an early Islamic site in the region (Thomas et al. 2006). Although far from ideal, these 'grab samples' from sections exposed by robber holes demonstrate the presence of a range of locally cultivable cereals, pulses and a variety of fruits at the site, as well as the expected dominance of sheep and goats amongst the faunal remains. The anthracological data and comparison of photographs from the Cambridge Expedition of 1959 with those from our fieldwork at Djām in 2005 concur with Helmut Freitag's assessment of the significant anthropogenic depletion of the region's vegetation, both in terms of extent and diversity (Deckers, in Thomas et al. 2006: 267–70, figs. 10–11; Freitag 1971: 95–7).

Other aspects of the region's environment have also been significantly altered, through a combination of natural and anthropogenic factors. The extensive irrigation canals to the south of G͟hūr along the Hilmand in Sīstān were part of a complex and well-maintained system designed to maximise the agricultural potential of the otherwise arid landscape—the contemporary chronicle, the *Tārīk͟h-i Sīstān*, refers to taxes spent preventing wind-driven sand from encroaching on agricultural land (Bosworth 2000: 38). Such large-scale environmental management projects were dependent upon highly skilled engineers and an ample supply of labour. The networks of dams, irrigation canals and subterranean water channels (*qanāt*s or *kārīz*) that were not destroyed by invading armies soon silted up due to a lack of maintenance.

The historical accounts and early Islamic ruins, which are scattered across much of Sīstān (Bosworth 2000; Fischer et al. 1974–76), are testimony to the region's erstwhile prosperity and subsequent decline in the face of the Mongol massacres, depopulation and mobile sands (Dupree 1977: 45). These examples illustrate that Braudel's concept of *structures* cannot be viewed as unchanging, nor can we simply transpose modern accounts of the region's physical environment onto the early Islamic period (*contra* Bowlby 1978: 13–17). They also highlight the lasting detrimental effect that depopulation can have on a region's infrastructure and managed environment.

HISTORICAL AND ARCHAEOLOGICAL INSIGHTS

The mobile subsistence strategies practised by much of the early Islamic population of central Asia are largely absent from the available archaeological record. The primary sources of archaeological information are derived from limited excavations at a few of the major urban centres. These hubs of commerce, culture and communication should be conceptualised as small islands of data linked by ephemeral networks, in a fluctuating and obscured sea of nomadic peoples, who are harder to define and trace archaeologically (Codrington 1944: 27; Rowton 1973). Adam T. Smith (2003: 109) describes the phenomenon as an 'archipelagic landscape.'

The G͟hūrids seem to have had a primarily subsistence-level economy until they rose to prominence as a coherent polity and military power during the first half of the twelfth century. Their armies sacked the G͟haznawid capitals of G͟hazna and Las͟hkar-i Bāzār/Bust in 1150–51 CE, and eventually eradicated the remnants of the G͟haznawid dynasty in the northern Indian subcontinent in 1186 CE (Bosworth 1977: 111–31). Despite their explosive entry onto the geo-political scene, the G͟hūrids struggled to build upon their initial military successes. The recapture of G͟hazna in 1173–74 CE after 12 years of the city being terrorised by "G͟huzz adventurers" (Bosworth 1965: 1102), however, proved pivotal. This victory enabled G͟hiyāt͟h al-Dīn, the greatest G͟hūrid sultan, to establish in effect a condominium with his brother Muᶜizz al-Dīn, whom he installed in G͟hazna. G͟hiyāt͟h al-Dīn campaigned in K͟hurāsān, while Muᶜizz al-Dīn used G͟hazna as a base from which to launch lucrative campaigns and territorial expansion into the northern Indian sub-continent. Significant but regrettably incalculable amounts of loot and trib-ute were transmitted from the northern Indian subcontinent to Fīrūzkūh via Ghazna (Juzjani 1970 I: 404–6), enabling Ghiyāth al-Dīn to aggrandise his summer capital and other centres, and the population of Fīrūzkūh to grow beyond the normal carrying capacity of the surrounding hinterland.

André Wink argues that the reason for the Ghūrids' military successes in the second half of the sixth/twelfth century was a fundamental change in the composition of their armies, with a shift from a predominance of local foot soldiers, practised in sieges of hilltop fortresses, to Turk cavalry regiments, better suited to engagements on the plains. It is notable that the Ghūrids' heterogeneous armies, which marched on the northern Indian subconti-nent in the late twelfth century, included significant numbers of Afghans, Tādjīks from Khurāsān and Khaldj troops from Zamīn-Dāwar (Wink 1991: 137). Although the details of this shift remain sketchy, the Ghūrids certainly gained access to a wider potential recruiting base around this time, as neighbouring amīrs and maliks in Khurāsān pledged their allegiance to Ghiyāth al-Dīn, and Ghuzz nomads penetrated the Ghazna region (Wink 1991: 137–8). Although the Ghuzz nomads are characterised as bellicose and destructive in the literary sources (Juzjani 1970 I: 376), they may have supplied horses for armies and slaves for the urban elites and the military, as other nomads did (Bosworth 1968: 5; Sinor 1990: 7–11).

THE GHŪRIDS' EXPANSION IN THE NORTHERN INDIAN SUBCONTINENT

A combination of religious zeal and looting appears to have motivated Muʿizz al-Dīn's campaigns in the northern Indian subcontinent (Anooshahr 2009: 5–6; Bosworth 1977: 8). Like their Ghaznawid predecessors, the Ghūrids made pragmatic accommodations with existing Indian polities. In Adjmēr, Dilhī (Delhi) and elsewhere, for example, local rulers or their off-spring accepted Ghūrid suzerainty and remained in power, in exchange for sending tribute to Ghazna. One facet of the Ghūrid campaigns that differs significantly from those of their predecessors, however, is in the creation of iqtāʿs, or grants of land from which the recipients could draw revenue (Lapi-dus 2002: 122–5). The physical manifestations of this imperial expansion are discussed by Patel (2009, 2012).

Muʿizz al-Dīn carefully controlled participation in the lucrative campaigns—only one Ghūrid prince is known to have accompanied him (Jackson 2000: 212). Iqtāʿs were assigned to Muʿizz al-Dīn's Turk mamlūks (slave soldiers), rather than to other members of the Ghūrid elite (Flood 2009: 112–4). Consequently, the Ghūrid elite based in central Afghanistan were unable to consolidate the dynasty's military victories and territorial gains in the northern Indian subcontinent after Muᶜizz al-Dīn's death and lost control of his *mamlūks*, whose primary allegiance was to their master rather than the Ghūrid polity.

TRANSFORMATIONS IN THE GHŪRID POLITY

Territorial expansion, both locally and into Khurāsān and the northern Indian subcontinent, inevitably affected the Ghūrid heartland in material, social and ideological terms. The historical sources suggest that the influx of tribute and loot injected an unprecedented amount of capital into the Ghūrid economy. Only traces of these 'new energy supplies' have survived, in the architectural remains and, to a lesser extent, material culture of the period (Flood 2009: 104; Morton 1978; Thomas 2010: 77–9). It is likely that much of the incoming wealth either consisted of, or was spent on, perishable goods to feed the seasonally swollen population at Fīrūzkūh.

Indic influences can be seen in the stone carvings on cenotaph fragments at Ghazna (Flood 2009: 190) and in the architecture of the Masdjid-i Sangī (Ball 1990). Magnificent structures such as the Minaret of Djām (Maricq & Wiet 1959; Sourdel-Thomine 2004), Shāh-i Mashhad *madrasa* (Glatzer 1973) and refurbished *masdjid-i djāmiᶜ* in Harāt (Hillenbrand 2002), however, clearly required significant levels of planning, investment, technical expertise and materials. The isolated locations of several of these monuments indicate the Ghūrids' transitory lifestyle.

Some of the incoming loot, such as golden trophies from Adjmēr, was displayed at Fīrūzkūh for propaganda purposes (Juzjani 1970 I: 404; Flood 2009: 126–7, 133), while the nouveau riche Ghūrid sultans became patrons of poets, religious scholars and intellectuals, and distributed largesse amongst their subjects (Juzjani 1970 I: 386–8, 405–7). The brief burgeoning of Fīrūzkūh's population and the presence of Jewish trading colonies and craftsmen, evident from the Hebrew-inscribed tombstones at Djām (Hunter 2010: 75; Lintz 2008, 2009), indicate other ways in which the appropriated wealth was spent.

The Ghūrids' florescence, however, was short-lived. As Ghiyāth al-Dīn succumbed to failing health, Muᶜizz al-Dīn spent more time campaigning in the west but could not prevent the loss of his brother's territorial gains in Khurāsān. When Muᶜizz al-Dīn was assassinated in 1206 CE, his Turk *mamlūks* took advantage of the ensuing vacuum to assert their power (Jackson 2000: 208, 211–16; Wink 1991: 140). The Ghūrid heartland and symbolically important throne in Fīrūzkūh mattered little to them; more crucially, the Turks possessed Muᶜizz al-Dīn's treasury, and had access to the northern Indian subcontinent, where the territorial gains of the previous decades remained unaffected by the resurgence of the Khʷārazm-Shāh in the west. Fīrūzkūh eventually fell to the Khʷārazm-Shāh in 1215–16 CE, and was reportedly destroyed after a second Mongol siege in 1222–23 CE (Juzjani 1970 I: 418–19; II: 1047–8). Although major urban centres such as Harāt eventually recovered, Fīrūzkūh was never re-occupied. The last members of the Ghūrid dynasty died as prisoners of the Kh^wārazm-Shāh or fled to Dilhī, where their *mamlūks* had established a series of short-lived Islamic dynasties collectively known as the Delhi Sultanate (1206–1527 CE).

CONCLUSIONS

The ultimate failure of the Ghūrids' imperial experiment reveals much about the intersection of environment, population and socioeconomic factors. Although not capable of sustaining large populations, the deserts and uplands of Afghanistan were an integral part of both local and regional economies, particularly for those with wide-ranging and flexible subsistence strategies. Local dynasts with broader ambitions, however, needed to look farther afield for the economic resources and manpower necessary to expand their realms. The predatory nature of these expansions made them unsustainable in the long term. The Ghūrid royal court required regular injections of loot and tribute to fund its building programmes and luxurious lifestyle; the rapid expansion of its empire overstretched the Ghūrids' armies, which became increasingly reliant upon mercenaries and slaves from the steppe, whose loyalties were fickle.

   The bloody demise of the Ghūrid dynasty (and their Ghaznawid predecessors) typifies the rapid rise and fall of what Bosworth refers to as 'power states,' reliant on military expansion and suppression for their survival (Bosworth 1998: 110; Sinopoli 1994: 162–4, 166–7). The political intrigue, factionalism (including the promotion of Turk mamlūks and other non-Ghūrid commanders) and the "unstructured and ad hoc cash dispensation through which the treasure obtained in the conquests was redistributed" (Wink 1991: 113–14) further destabilised the empire. The longevity of Islamic dominion on the Indo-Gangetic plains indicates that the Ghūrid expansion eastwards beyond their heartland was not a complete failure. Ironically, however, it was the Turk *mamlūks* who benefited rather than their Ghūrid masters, who were too ideologically attached to their heartland in the mountains of central Afghanistan. In the face of military defeats and an economic crisis, the Ghūrids downsized and reverted to their localised, transitory way of life, which can still be seen today.

ACKNOWLEDGMENTS

I am grateful to my father, Colin Thomas, for a lifetime of musings on human geography, particularly population movements, and to Fiona Kidd and Alka Patel for their pertinent comments on previous drafts of this chapter.

NOTE

Transliteration and diacritics follow, in the main, the *Encyclopaedia of Islam* (Leiden: E.J. Brill). The definition of 'early Islamic' in Afghanistan as dating from about 622–1220 CE follows Ball (2008: 88–94).

REFERENCES

Anon. 1970. *Hudūd al–cĀlam 'The regions of the world.' A Persian geography 372 A.H.–982 A.D. Translated and explained by V. Minorsky*. London: Luzac.

Anooshahr, A. 2009. *The Ghazi sultans and the frontiers of Islam: A comparative study of the Late Medieval and Early Modern periods*. London: Routledge.

Ball, W. 1990. Some notes on the Masjid–i Sangi at Larwand in central Afghanistan. *South Asian Studie*s 6: 105–10.

Ball, W. 2002. The towers of Ghur: a Ghurid Maginot line?, in W. Ball & L. Harrow (ed.) *Cairo to Kabul. Afghan and Islamic studies presented to Ralph Pinder-Wilson*: 21–45. London: Melisende.

Ball, W. 2008. The *monuments of Afghanistan. History, archaeology and architecture*. London: I.B. Tauris.

Bintliff, J. 2006. Time, structure and agency: the Annales, emergent complexity, and archaeology, in J. Bintliff (ed.) *A companion to archaeology*: 174–94. Oxford: Blackwell.

Bosworth, C.E. 1961. The Early Islamic history of Ghur. *Central Asiatic Journal 5*: 116–33.

Bosworth, C.E. 1965. Ghurids, in E. van Donzel, B. Lewis & C. Pellat (ed.) *The encylopaedia of Islam* II: 1099–104. Leiden: E.J. Brill.

Bosworth, C.E. 1968. The political and dynastic history of the Iranian world (A.D. 1000–1217), in J.A. Boyle (ed.) *The Cambridge history of Iran. Volume V: The Saljuq and Mongol periods*: 1–202. Cambridge: Cambridge University Press.

Bosworth, C.E. 1977. *The later Ghaznavids: splendour and decay. The dynasty in Afghanistan and northern India 1040–1186*. Edinburgh: Edinburgh University Press.

Bosworth, C.E. 1998. The Ghaznavids, in M.S. Asimov & C.E. Bosworth (ed.) *History of civilizations of central Asia. Vol. IV. The age of achievement: A.D. 750 to the end of the fifteenth century. Part one. The historical, social and economic setting*: 95–117. Paris: UNESCO Publishing.

Bosworth, C.E. 2000. Sistan and its local histories. *Iranian Studies* 33(1): 31–43.

Bowlby, S.R. 1978. The geographical background, in F.R. Allchin & N. Hammond (ed.) *The archaeology of Afghanistan from earliest times to the Timurid Period*: 9–36. London: Academic Press.

Braudel, F. 1972. *The Mediterranean and the Mediterranean world in the age of Philip II*. London: Collins.

Brown, L.R. 2005. *Outgrowing the earth: the food security challenge in an age of falling water tables and rising temperatures*. New York: W.W. Norton.

Chase-Dunn, C. & T.D. Hall. 1993. Comparing world-systems: concepts and working hypotheses. *Social Forces* 71(4): 851–86.

Codrington, K.d.B. 1944. A geographical introduction to the history of Central Asia. *The Geographical Journal* 104(½): 27–40.

Cribb, R. 2004. *Nomads in archaeology*. Cambridge: Cambridge University Press.

Dupree, N.H. 1977. *An historical guide to Afghanistan*. Kabul: Afghan Tourist Organization.

Fischer, K., D. Morgenstern & V. Thewalt (ed.). 1974–76. *Nimruz. geländebegehungen in Seistan 1955–1973 und die aufnahme von Dewal–i Khodaydad 1970.* Bonn: Rudolf Habelt Verlag.

Flood, F.B. 2009. *Objects of translation: artifacts, elites and Medieval Hindu–Muslim encounters.* Princeton: Princeton University Press.

Freitag, F.B. 1971. Studies on the natural vegetation of Afghanistan, in P.H. Davis, P.C. Harper & I.C. Hedge (ed.) *Plant life of south-west Asia*: 89–106. Edinburgh: Royal Botanic Gardens.

Glatzer, B. 1973. The madrasah of Shah–i–Mashhad in Bagdis. *Afghanistan* 25(4): 46–68.

Hillenbrand, R. 2002. The Ghurid tomb at Herat, in W. Ball & L. Harrow (ed.) *Cairo to Kabul. Afghan and Islamic studies presented to Ralph Pinder-Wilson*: 123–43. London: Melisende.

Horden, P. & N. Purcell. 2000. *The corrupting sea: a study of Mediterranean history.* Malden: Blackwell.

Hunter, E.C.D. 2010. Hebrew-script tombstones from Jām, Afghanistan, *Journal of Jewish Studies* 61(1): 72–87.

Jackson, P. 2000. The fall of the Ghurid dynasty, in C. Hillenbrand (ed.) *Studies in honour of Clifford Edmund Bosworth volume II. The sultan's turret: studies in Persian and Turkish culture*: 207–35. Leiden: Brill.

Juzjani, 1970 [1881]. *Tabakāt–i–Nāsirī: a general history of the Muhammadan dynasties of Asia, including Hindustan; from A.H. 194 (810 A.D.) to A.H. 658 (1260 A.D.) and the irruption of the infidel Mughals into Islam.* New Delhi: H.G. Raverty (Oriental Books Reprint Corporation).

Kohzad, A.A. 1957. Firoz Koh. *Afghanistan* 12(4): 31–4.

Kramer, C. 1982. *Village Ethnoarchaeology: rural Iran in archaeological perspective.* New York: Academic Press.

Kremenetski, K.V., T. Boettger, G.M. Macdonald, T. Vaschalova, L. Sulerzhitsky & A. Hiller. 2004. Medieval climate warming and aridity as indicated by multiproxy evidence from the Kola Peninsula, Russia. *Palaeogeography, Palaeoclimatology, Palaeoecology* 209: 113–25.

Lapidus, I.M. 2002. *A history of Islamic societies.* Cambridge: Cambridge University Press.

Le Roy Ladurie, E. 1974. *The peasants of Languedoc.* Urbana: University of Illinois Press.

Le Roy Ladurie, E. 1978. *Montaillou. Cathars and Catholics in a French village 1294–1324.* London: Scolar Press.

Le Strange, G. 1976 [1905]. *The lands of the eastern Caliphate. Mesopotamia, Persia and Central Asia from the Moslem conquest to the time of Timur.* New York: AMS Press.

Lewthwaite, J. 1988. Trial by durée: a review of historical-geographical concepts relevant to the archaeology of settlement on Corsica and Sardinia, in J.L. Bintliff, D.A. Davidson & E.G. Grant (ed.) *Conceptual issues in environmental archaeology*: 161–86. Edinburgh: Edinburgh University Press.

Lintz, U.-C. 2008. Persisch-Hebräische Inschriften aus Afghanistan. *Judaica* 64(4): 333–58.

Lintz, U.-C. 2009. Persisch–Hebräische inschriften aus Afghanistan (Teil II). *Judaica* 65(1): 43–74.

Maricq, A. & G. Wiet. 1959. *Le minaret de Djam: la découverte de la capitale des sultans Ghurides (XIIe–XIIIe siècles).* Paris: Mémoires de la délégation archéologique française en Afghanistan.

Mohammed, A. 1973. The nomadic and the sedentary: polar complementaries—not polar opposites, in C. Nelson (ed.) *The desert and the sown: nomads in the wider society. Papers presented at a conference held in March 1972 at the American*

University in Cairo: 97–112. Berkeley: Institute of International Studies, University of California.

Morton, A.H. 1978. Ghūrid gold en route to England? *Iran* 16: 167–70.

Patel, A. 2009. Expanding the Ghurid architectural corpus east of the Indus: the Jagesvara temple at Sadadi, Rajasthan. *Archives of Asian Art* 59: 33–56.

Patel, A. 2012. Architectural cultures and empire: the Ghurids in northern India (ca. 1192–1210). *Bulletin of the Asia Institute* 21: 35–60.

Rowton, M.B. 1973. Urban autonomy in a nomadic environment. *Journal of Near Eastern Studies* 32(½): 201–15.

Sanderson, S.K. 1988. What kind of ecologist was Braudel? A comment on Hudson. *Sociological Forum* 3(2): 277–80.

Sinopoli, C.M. 1994. The archaeology of empires. *Annual Review of Anthropology* 23: 159–80.

Sinor, D. 1990. Introduction: the concept of Inner Asia, in D. Sinor (ed.) *The Cambridge history of early Inner Asia*: 1–18. Cambridge: Cambridge University Press.

Smith, A.T. 2003. *The political landscape. Constellations of authority in early complex polities.* Berkeley: University of California Press.

Sourdel-Thomine, J. 2004. *Le minaret Ghouride de Jām. Un chef d'œuvre du XIIe Siècle.* Paris: de Boccard.

Sumner, W.M. 1989. Population and settlement area: an example from Iran. *American Anthropologist* (NS) 91(3):631–41.

Taaffe, R.N. 1990. The geographic setting, in D. Sinor (ed.) *The Cambridge history of early Inner Asia*: 19–40. Cambridge: Cambridge University Press.

Taagepera, R. 1978. Size and duration of empires: systematics of size. *Social Science Research* 7: 108–27.

Thomas, D.C. 2007. Firûzkûh: the summer capital of the Ghurids, in A.K. Bennison & A.L. Gascoigne (ed.) *Cities in the pre-modern Islamic world: the urban impact of state, society and religion*: 115–44. London: Routledge Curzon.

Thomas, D.C. 2010. Signifying the 'Ghūrid self'—the juxtaposition between historical labels and the expression of identity in material culture. *Journal of Historical and European Studies* 3: 71–88.

Thomas, D.C., K. Deckers, M.M. Hald, M. Holmes, M. Madella & K. White. 2006. Environmental evidence from the Minaret of Jam Archaeological Project, Afghanistan. *Iran* 44: 253–76.

Trouet, V., J. Esper, N.E. Graham, A. Baker, J.D. Scourse & D.C. Frank. 2009. Persistent positive North Atlantic Oscillation mode dominated the Medieval Climate Anomaly. *Science* 324(5923): 78–80.

Wallerstein, I. 1976. *The modern world-system. Vol. 1: capitalist agriculture and the origins of the European world-economy in the sixteenth century.* New York: Academic Press.

Wheatley, P. 2001. *The places where men pray together: cities in Islamic lands.* Chicago: University of Chicago Press.

Wink, A. 1991. *Al-Hind: the making of the Indo-Islamic world, Volume 2: the slave kings and the Islamic conquest, 11th–13th centuries.* Leiden: E.J. Brill.

Part II

Technology and the Environment

7 A Long-Term History of Horticultural Innovation and Introduction in the Highlands of Papua New Guinea

Tim Denham

INTRODUCTION

People in the interior of New Guinea traditionally lived the majority of their lives within a relatively circumscribed territory. Their world was a densely structured landscape imbued with meaning (e.g. Weiner 1991). This textual richness was polyvalent and multilayered; overlapping and interweaving senses bound people to place. However, in trying to understand what people were doing in these landscapes in the past, we are limited by the physical evidence of former practices. Try as we might, we are unable to recover the specific realms of meaning that structured their lives; such attempts are fraught and liable either to fall into overinterpretation (Eco 1991) or rely too heavily on ethnographic analogy (Wylie 1985). As a result, we are left trying to document 'how,' rather than 'why,' people interacted with their world.

In this chapter, the historical development of practices associated with the exploitation of plants for food is charted for a landscape in the Upper Wahgi Valley, Papua New Guinea. It focuses upon a series of technological innovations and introductions from the beginning of the Holocene, about 11,500 years ago, up to the recent past. Once adopted, these innovations and introductions transformed and broadened the way people engaged with their environment, as well as with each other. The chapter concludes with a consideration of how an understanding of the past has relevance for understanding the resilience of traditional forms of horticulture in the highlands of Papua New Guinea today.

TECHNOLOGY AND PRACTICE

Technology represents an orientation to the world that is shared to varying degrees by members of a particular group or society. Technology does not just refer to how people interact with their material world, but how people interact with the world as it is meaningfully constituted. Whether or not people use a particular technology does not alter the fact that they are emplaced within a society characterised by an ever-changing, polythetic suite of varyingly deployed technologies. In Papua New Guinea this can be exemplified in the present by the diverse forms of agriculture that draw on relatively common sets of cultivation practices and plants (Sillitoe 1989; Bourke & Harwood 2009).

In an everyday sense, technology is concerned with the practical. People use various technologies as a means of interacting with their world. Technologies exist at the level of historical actuality and can be described as they were used in particular times and places. However, technologies have immaterial and material components. For each technology there is the knowledge of how to do something and there is the actual doing of something. Technologies also persist through time: they are temporal.

From this perspective, technology is viewed in its instrumentality, as a mechanism through which people achieve mastery over things, including of their environment and of other people. Given that technologies do not exist beyond the realm of human experience, they can be considered to exist through practice. Practice refers to the habitual activities of people, whether they are a product of structuring influences, dispositions or individual whim (Bourdieu 1990): practices are what people do. As such, practices emplace technologies. From an archaeological perspective, technologies only exist through evidence of the practices through which they were deployed.

A PRACTICE-CENTRED METHOD

A practice-centred method has been designed to develop a common conceptual basis and language to discuss all forms of plant exploitation in the past or present (Denham 2008a, 2008b, 2009; Denham et al. 2009a; cf. Bourke 2001). This conceptual framework is comparable to the skills-based provisioning spreadsheet proposed by Latinis (2000; Terrell et al. 2003), although that was focused on resource exploitation in agroforestry. The intention is to free debate from the seemingly ever-present problems of differentiating foragers or hunter–gatherers from farmers. Numerous working definitions have been proposed to distinguish these conceptual categories and to clarify the transitional 'middle ground' between them (Harris 1989; Smith 2001; Roscoe 2002), or to radically overthrow the whole agenda (Terrell et al. 2003).

There is a growing acceptance that forms of food production in different parts of the world may need a contingent, or more malleable, conceptual framework for classification (Denham 2007a, 2011). Researchers in different regions rely on diverse lines of evidence to define and differentiate agriculture from other activities in the past (compare papers in Harris 1996 to those in Denham et al. 2007). Within tropical forest locations, most authors recognise that the conceptual demarcations between traditional forms of plant domestication, agriculture and horticulture are unclear and porous (e.g. Piperno & Pearsall 1998).

Domestication is ordinarily defined as the accumulation of phenotypic and genotypic traits derived from the long-term human selection of an animal or plant. However, there are several debates that cast doubt on the clarity of any definition of domestication: the relatively arbitrary differentiation between management and domestication (Yen 1989), the relative importance of human selection versus genetic isolation (Jones & Brown 2007), the relative importance of the social and environmental contexts of domestication for the emergence and nature of phenotypic traits (Marshall 2007; Pearsall 2007), and the role of gene expression and phenotypic plasticity in blurring the putative correlation between genotypic and phenotypic traits (Gremillion & Piperno 2009).

In New Guinea the identification of agriculture in the past is determined by a reliance on the cultivation of plants for food, irrespective of domesticated status (Denham 2007a; cf. Spriggs 1996). Such reliance is more visible in highly altered anthropic landscapes, such as montane rainforests denuded to grasslands (Haberle 2003). These environments are depauperate in other edible resources and so people were increasingly dependent on growing their own food.

Horticulture is considered to be a subset of agriculture that is characterised by the mixed cropping, or intercropping, of plants in prepared plots. In the New Guinea context, horticultural plots contain an admixture of plants with varying degrees of domestication (Denham 2005a; cf. Caballero 2004). Many greens are propagated directly from wild stock, hence a lack of domesticated traits, whereas most staples are propagated from cultivated stock, hence an accumulation of domesticated traits, albeit diluted when inbreeding occurs with wild populations (see Kennedy & Clarke 2004).

The agricultural chronology from the Upper Wahgi Valley is not reliant on the presence or abundance of domesticated plants. There is evidence for the exploitation of yam (*Dioscorea* sp.) and taro (*Colocasia esculenta*) from *c*. 10,000 cal BP and the cultivation of bananas (*Musa* sp.) from 6950/6440 cal BP (Denham et al. 2003). However, despite genetic evidence suggesting the domestication of these and other crop plants in the New Guinea region (Lebot 1999), there is currently a lack of corresponding archaeobotanical evidence documenting the sequential domestication of any plant (Denham 2011; *contra* Yen 1996). Indeed, many staples are still considered to be semidomesticated (Yen 1991) and numerous other plants, including greens, are vegetatively propagated from wild or feral stands. A marked feature of agricultural and arboricultural practices in New Guinea is the vegetative propagation of most plants and the transplantation of trees using seedlings, respectively. These clonal reproductive practices need not produce the same diagnostic transformations characterised as the 'domestication syndrome' in plants cultivated for their seeds (Gepts & Papa 2002).

A CHRONOLOGY OF PRACTICES

Multidisciplinary evidence can be used to reconstruct a chronology of practices that are constituent for multiple forms of plant exploitation in the Upper Wahgi Valley (Denham & Haberle 2008; Denham 2009; Denham et al. 2009a). Findings from several sites within one landscape provide complementary evidence of plant exploitation across the landscape from the Pleistocene to Late Holocene (Figure 7.1). Investigations include:

- archaeological excavations in wetlands revealing evidence of topographical manipulation, mounding and ditched drainage (Golson 1977; Denham et al. 2004a)
- palaeoecological reconstructions of wetlands at Kuk, Warrawau, Lake Ambra and Ambra Crater, as well as at Manim rockshelter, that collectively show environmental changes over the last 30,000 years (Powell 1982a; Denham & Haberle 2008; Sniderman et al. 2009)
- geomorphological investigations of slope erosion/deposition rates at Kuk and Manim, respectively (Hughes et al. 1991)
- archaeological excavations of rockshelters along an altitudinal gradient in a tributary valley (Wurup Valley)—Manim (1770m), Kamapuk (2050m), Etpiti (2200m) and Tugeri (2450m) (Christensen 1975)

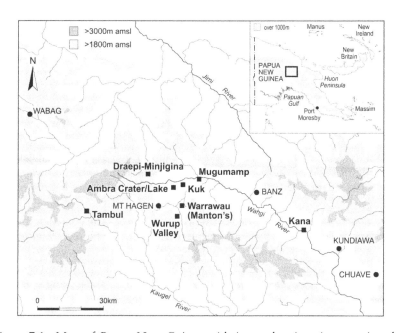

Figure 7.1 Map of Papua New Guinea with inset, showing sites mentioned in the text

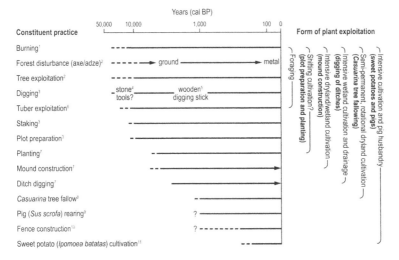

Figure 7.2 Chronology of practices and forms of plant exploitation in the Upper Wahgi Valley (amended version of Denham 2009: fig. 1)

Notes: [1]Denham et al. 2004b; [2]Christensen 1975; [3]Denham 2005b; [4]Bulmer 2005; [5]Golson et al. 1967; [6]Fullagar et al. 2006; [7]Denham et al. 2003; [8]Haberle 2007; [9]Sutton et al. 2009; [10]Golson 1977; [11]Golson 1982

The antiquity and temporal extension of each practice has been reconstructed using a combination of archaeobotany (plant use), archaeology (cultivation practices) and palaeoecology (environmental transformation) (Denham & Haberle 2008) (Figure 7.2). For example, anthropic burning is inferred from elevated charcoal frequencies and prolonged decreases in primary forest taxa with concomitant increases in secondary forest taxa and grasslands (Haberle 1994; Hope 2009). The historical and geographical specificity of each line of evidence is secure because each has been dated and most are derived from a single landscape.

Once the chronology has been established, the practices are 'bundled' in time and space to construct different forms of plant exploitation (Denham 2009). The charting of individual practices and 'bundling' them in space and time draws explicitly on the time geography of Hägerstrand (1970; also see Gregory 2000). A focus on one landscape rather than on a larger geographical region avoids problems of decontextualising practices in sociospatial terms, thereby resulting in the conflation and bundling of practices that never co-occurred in the past.

INNOVATIONS AND INTRODUCTIONS

The sequential history of agricultural emergence and transformation in the Upper Wahgi Valley, and elsewhere in the highlands, has formerly relied on

Golson's phases for Kuk and resultant inter-site comparison (Golson 1977, 1982; Denham 2003a, 2005a, 2007a; Bayliss-Smith 2007). Here, the use of chronological phases as an organising principle for discussion is abandoned in favour of viewing agricultural history as a series of innovations and introductions. This shift is justified because constituent practices and forms of plant exploitation do not occur just within a set period, only to then be abandoned; rather, they persist through time and are still practised by people living in the valley or adjacent regions today (Figure 7.3; Bourke & Harwood 2009). The antiquity, timing of widespread adoption and resultant implications of major technological innovations and introductions—documented through various practices—are significant for understanding the long-term history of horticulture in the highlands. It is worth noting, too, that the timing of innovation and introduction may be much earlier than the timing of widespread adoption.

The articulation of constituent practices into different forms of plant exploitation should not be inferred to represent a developmental trajectory. If the Upper Wahgi Valley is considered in isolation, or if New Guinea is considered as a whole, multiple forms of plant exploitation occurred at the same time in different places, as they still do. Different forms of plant exploitation should be viewed as an expanding repertoire deployed by people in different ways in different social contexts. More forms can be identified towards the present, yet some unidentified forms may no longer occur, while others are marginal or have been abandoned (Terrell 2002; cf. Balée 1994). Although theoretically it may be possible to identify novel or extinct forms of plant exploitation, namely those that occurred in the past for which there are no contemporary analogues, the multidisciplinary evidence is insufficiently robust to do so.

The interpretation of forms of plant exploitation from the temporal bundling of constituent practices is not just reliant on co-occurrence. It is necessary to have some interpretative insights. For example, the differentiation of the earliest cultivation, as opposed to wild plant resource exploitation or intensification (Gott 2005), focuses on bananas (*Musa* spp., based on anomalously high frequencies of banana phytoliths; Denham et al. 2003), the bases of former mounds (excavated archaeologically; Denham et al.

Figure 7.3 Composite image showing practices discussed in the text: A. Karuka or highland pandanus (*Pandanus brosimos/iwen/jiulianettii*) (Photo: Alice Sutton, Simbai, 2007); B. Newly planted swidden plot (Photo: Tim Denham, Lower Jimi Valley, 1990); C. Forest-grassland mosaic (Photo: Tim Denham, Bismarck Mountain Range, 1990); D. Chopping down a tree with a stone adze (Photo: Peter White, Hewa, 1967); E. Gardens planted with mounds of sweet potato (*Ipomoea batatas*) (Photo: Tim Denham, Southern Highlands Province, 2007); F. Dryland ditch on a hill slope (Photo: Alice Sutton, Simbai, 2007); G. Current garden in front of an old garden plot planted with *Casuarina oligodon* tree-fallow (Photo: Alice Sutton, Simbai, 2007); H. Tethered pigs (*Sus scrofa*) (Photo: Tim Denham, Mount Hagen vicinity, 1990)

2004a) and major environmental transformations, including the degradation of montane forest on the valley floor to grassland (from microcharcoal and pollen records; Haberle et al. 2102). It is not the presence or necessarily types of banana that are significant, but the abundance that is suggestive of cultivation.

The innovations discussed below are all considered to have been independently developed by New Guineans. The exceptions are the bookends: foraging practices were shared by the colonists of Sunda and Sahul (Denham et al. 2009a), and intensive sweet potato cultivation (*Ipomoea batatas*) and domestic pig rearing (*Sus* sp.) are the products of two relatively recent introductions. For each independent innovation, the locus of inception within the New Guinea region is not known. Although the earliest evidence for most activities derives from the robust multidisciplinary record in the Upper Wahgi Valley, this is almost certainly a sampling effect, since this area has been investigated in greatest detail.

Foraging

Foraging is characterised by the exploitation and management of resources, including through burning, gap maintenance and the creation of patches within the rainforest (Denham & Barton 2006). These activities are designed to increase the density of resources within the landscape by increasing the number and productivity of favoured plants (Yen 1989). These practices, including ring-barking using waisted blades, are hypothesised to have occurred since human colonisation of New Guinea sometime before 45,000 years ago (Groube 1989).

The earliest evidence of anthropic burning and forest disturbance in the Upper Wahgi Valley is approximately 30,000 years old (Denham & Haberle 2008). These dates correspond to those of the earliest human activity on the wetland margin at Kuk, represented by a possible 'hearth' dated to 36,520–32,240 cal BP (30,000 ± 950 BP, ANU 3188; Hughes et al. in press; Hope & Haberle 2005). Although Pleistocene-aged axes/adzes have not been found in the Upper Wahgi Valley, they do occur elsewhere in the highlands (Mountain 1991; Summerhayes et al. 2010) and lowlands (Groube et al. 1986).

The exploitation of highland environments during the Late Pleistocene, especially the Last Glacial Maximum, has recurrently been portrayed as focused on the seasonal exploitation of the high altitude *Pandanus brosimos/iwen/jiulianettii* complex (Golson 1991). However, the floors of many intermontane valleys retained a more diverse abundance of faunal and floral resources throughout the period of human occupation (Denham 2007b). Based on the limited evidence, human occupation on the floor of the Upper Wahgi Valley during the Pleistocene has been characterised as broad-spectrum foraging (Haberle et al. 2012: 136; also see Denham & Barton 2006 and Denham 2007b):

Some gaps in the forest were maintained through burning and clearing, and patches of grassland formed, potentially adjacent to wetlands and along riparian corridors, due to localized and sustained forest disturbance. As patches became maintained foci of activity, so too the resources within those gaps—including herbs (*Musa* spp.), tuberous plants (potentially including taro and yams), grasses (*Saccharum* spp. and *Setaria palmifolia*), and a wide variety of leafy vegetables—were brought under increasing management. At this time, people are inferred to have engaged in extensive hunting and foraging activities to sustain broad-spectrum diets. (Haberle et al. 2012: 136)

The earliest evidence for digging and the exploitation of tuberous plants at Kuk (Fullagar et al. 2006), as well as the exploitation of nut-bearing species (*Pandanus*) at Manim (Christensen 1975), dates to the Pleistocene/Holocene transition. These activities are only inferred for the Upper Wahgi Valley based on generalised patterns of human behaviour in Sahul (Denham et al. 2009a) and through comparison with other highland and lowland locales (Fairbairn 2005; Summerhayes et al. 2010).

Foraging activities continue to be practised by people across New Guinea. Wherever people have access to areas of primary and secondary forest, or even grassland, they exploit a variety of plant (and animal) resources for food, construction, decoration, medicine and so on (Powell 1982a). People in the highly altered and cultivated landscapes of the Upper Wahgi Valley today often have to travel to neighbouring valleys to engage in foraging and hunting.

Shifting Cultivation (Innovations: Plot Preparation and Planting)

Shifting cultivation entails the clearing of a plot of land for cultivation for a short period with subsequent abandonment to fallow for a longer period (Clarke 1971). Clearing usually occurs using a slash-and-burn technique, or in areas with perennially high rainfall a slash-and-mulch variant. Both are designed to release nutrients from the vegetation to aid development of planted crops. The cleared plot is cultivated continuously for periods varying from one to several years, although some cultivated plants and tree crops will be harvested and managed for many subsequent years. Following the abandonment of active cultivation, the plot is usually left fallow for periods upwards of five years.

A combination of practices suggests that shifting cultivation may have been practised on the floor of the Upper Wahgi Valley during the Early Holocene (Denham 2005b; Denham & Barton 2006; Denham & Haberle 2008). Multiple lines of evidence from a relatively restricted area at Kuk are suggestive of swidden cultivation at *c.* 10,000 cal BP (Denham 2005b). These include:

- a vegetation mosaic of primary forest, secondary forest and grassland on the valley floor, representing the progressive opening up of the montane forest environment using fire (Haberle et al. 2012)
- evidence of localised clearance in palaeochannel fills (Denham et al. 2009b), increased erosion rates within the catchment and increased deposition of an alluvial fan on the wetland margin (Hughes et al. 1991)
- manipulation of the wetland margin, including modification of surfaces to aid drainage, digging of pits to harvest and possibly plant tubers, and the use of stakes to harvest (e.g. bananas) or support (e.g. edible cane grasses—*Saccharum* and *Setaria*) plants (Denham et al. 2004a)
- targeted procurement, processing and consumption of starch-rich edible plants, including a yam and taro (Fullagar et al. 2006)

These lines of evidence are not definitive, primarily because the archaeological visibility of minimal tillage techniques characteristic of swidden plots is low and because there is no proof of planting. The limited potential for finding archaeological remains of swiddening activities of any antiquity requires a reliance on inferential evidence: palaeoecology of vegetation change (Powell 1982a) and sedimentology of soil profiles such as *terra mulata* in Amazonia (Arroyo-Kalin 2008). The presence (bananas) and exploitation (taro and a yam) of food plants potentially of lowland derivation is significant. If these plants are of ultimate lowland derivation (as suggested by Yen 1995) and were brought into the highlands by people, then they must have presumably been transplanted there as part of some form of cultivation practice. However, caution is needed:

> The appearance of these starch rich plants at Kuk at the beginning of the Holocene may relate to natural dispersal processes (i.e. seed-dispersed under warming climate conditions) as much as to human agency (i.e. brought by mobile groups engaged in an extensive form of plant exploitation based on vegetative propagation). (Denham et al. 2004b: 852)

Even if these plants were found wild in the Upper Wahgi Valley, they were at the edge of their range during the Pleistocene and Holocene. Therefore, they are unlikely to reproduce sexually, suggesting a human role in their subsequent vegetative dispersal. In sum, Early Holocene practices plausibly represent swidden cultivation, or a transitional form of patch-based foraging (Denham & Haberle 2008). Palaeoecological and geomorphological evidence suggesting similar practices persists throughout the Early Holocene at Kuk, even though the wetland archaeological record of these activities is not preserved.

Shifting cultivation is a widespread practice in the fringe highlands and lowlands of New Guinea today. Although Denevan (2001) has suggested that shifting cultivation was not possible within rainforest environments in the lowland neotropics before metal tools, this scenario does not apply to New Guinea. Ethnographic and ethnoarchaeological accounts document traditional shifting cultivation using stone tools within rainforest environments across the island (e.g. Pétrequin & Pétrequin 2006). Similarly, montane rainforest environments across the island show degradation to grassland throughout the Holocene, although these effects are generally not progressive before around 4000 years ago, except in the Upper Baliem and Upper Wahgi valleys (Hope & Haberle 2005). In the New Guinea context, it is more appropriate to say that shifting cultivation was greatly facilitated by the emergence of highly ground axes/adzes from *c.* 7000 years ago (Christensen 1975).

Intensive Dryland/Wetland Margin Cultivation (Innovation: Mound Construction)

Modern forms of intensive dryland and wetland cultivation across New Guinea use various types of mounds or raised beds (Serpenti 1965; Waddell 1972; Bourke & Harwood 2009; Hitchcock 2010). These kinds of agricultural intensification constitute a form of tillage that is designed, often in combination with green or ash composting, to prolong the fertility of the cultivated plot (Powell et al. 1975; Bourke 2001). The agronomic benefits of cultivating in mounds (of varying sizes) are well documented across highland provinces (e.g. Brookfield & Brown 1963; Waddell 1972).

At 6950/6440 cal BP there is clear evidence for cultivation using mounds on the wetland margin at Kuk (Denham et al. 2003) (Figure 7.4), with later comparable evidence at two other locations in the Upper Wahgi Valley (Denham 2003a, 2007a): Mugumamp (Harris & Hughes 1978) and Warrawau (Golson 1982, 2002). At Kuk the curvilinear bases of cultivation mounds are associated with banana cultivation; continued exploitation of other plants, including an aroid; and a dramatic replacement of forest with grassland on the valley floor (Denham et al. 2003, 2009b; Denham & Haberle 2008; Haberle et al. 2012; cf. Sniderman et al. 2009). These practices are consistent with garden cultivation using mounds, which occurred in some form or another across the valley floor.

Although it has been argued that mound cultivation was originally a wetland adaptation (Golson 1977; cf. Leahy 1936), it seems equally plausible that cultivation using mounds occurred on valley slopes as well as on wetland margins. On wetland margins mounds would enable the cultivation of plants with different edaphic requirements: water-tolerant plants (e.g. taro) could be grown in the wetter soils around the base of the mounds, while water-intolerant plants (e.g. sugarcane) could be grown on the drier mounds (following Golson 1977). In the past, mound cultivation plausibly occurred across the landscape, but archaeological evidence has only been preserved in the wetlands. Despite these uncertainties, mounds have persisted as a technological practice from at least 6950/6440 years ago to the present.

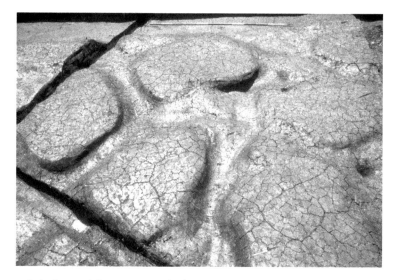

Figure 7.4   Base of former mounds dating to 6950/6440 cal BP at Kuk (Photo: Alistair Marshall, Kuk, 1976; courtesy Jack Golson). The ranging pole is 2.5m

Intensive Wetland Cultivation and Drainage (Innovation: Digging of Ditches)

Curvilinear and rectilinear drainage networks comprised of ditches exist in many highland and lowland regions across New Guinea today. In the highlands, ditches are ordinarily associated with wetland drainage for cultivation. However, ditches have multiple functions and in many areas are dug across valley slopes for defence, to demarcate clan boundaries, to enclose or exclude pigs, and to define sacred or spiritual places (Ballard 1995, 2001; author's personal observations from fieldwork, 1990–2010). In the absence of domesticated pigs in the highlands before about 2000 years ago, ditches in wetlands were primarily dug to define and drain plots for the cultivation of plants.

Although ditch networks require relatively high labour inputs to construct (Steensberg 1980), the enhanced fertility of soils in wetlands enables greater yields and continuity of cultivation. Moreover, the availability of soil water for crops in wetlands ensures more reliable yields during times of climatic stress, such as the increased frequency of droughts in the highlands following intensification of El Niño–Southern Oscillation (ENSO) events during the mid- to late Holocene (Moy et al. 2002). Because of the combined benefits of higher yields and risk reduction, wetland plots have become increasingly important and reliable foci of agricultural production.

The earliest ditch networks in the Upper Wahgi Valley date from around 4400/4000 cal BP (Figure 7.5; Denham 2005a). The antiquity of the earliest ditches corresponds to the age of the earliest wooden digging implement—a

Figure 7.5 Multiple ditches exposed in the base of an excavation at Kuk (Photo: Tim Denham, Kuk, 1999). The ranging pole is 1m

hastate-type wooden spade used in ditch maintenance—from Tambul in the neighbouring Upper Kaugel Valley (Golson 1997). Ditch networks become more ubiquitous at wetlands within the valley from around 2300 years ago (Kuk, Minjigina and Warrawau; Denham 2005a), as at other intermontane wetlands (Ballard 1995; Muke & Mandui 2003). A wooden digging stick was excavated at Warrawau in 1966 and dated to 2300 ± 120 BP (ANU 43; Golson et al. 1967). Although well defined only at Kuk, the designs of these relatively early ditch networks are highly variable, including dendritic, rectilinear and trellis-like forms (Denham 2005a).

In successive drainage networks the alignments and size of enclosed plots varies, but all are based on a grid design (Bayliss-Smith 2007). Ditches demarcate quadrangular plots used for cultivation. Numerous digging sticks and wooden spades comparable to those documented ethnographically have been collected from archaeological contexts dating to the last 2000 years (Golson & Steensberg 1985).

Golson (1977) has proposed that ditch networks were periodically dug in wetlands in response to crises in food production in the dryland sphere (cf. Gorecki 1979). These disruptions can be varyingly attributed to soil nutrient depletion, overpopulation, climatic stresses and so on (Bayliss-Smith 2007). However, drained plots in wetlands have formed part of agricultural practices in the valley relatively continuously from at least 2300 years ago. The apparent periodic abandonment and redigging of ditch networks reflects the spatially variable usage of wetlands and areas within them (following Ballard 2001), even though some wetlands may have been episodically abandoned following warfare, as recorded in ethnohistories for Kuk (Ketan 1998).

Semipermanent, Rotational Dryland Cultivation
(Innovation: *Casuarina* Tree-Fallowing)

Tree planting in fallows is a practice used by highlanders to 'fix' nitrogen and increase carbon in the soil, thereby enriching it for subsequent cultivation (Bourke & Harwood 2009: 245–7). The most common practice is to transplant seedlings of *Casuarina oligodon* into plots near the end of the cropping cycle. These trees grow for about 8–12 years before being cut down, ring-barked or pollarded prior to subsequent cultivation of the plot. In addition to being important for improving soils, fast-growing *Casuarina* is an important source of firewood and timber for construction.

Although *Casuarina* has been a feature of the valley's flora from at least the beginning of the Holocene (Haberle et al. 2012), Haberle (2007) has inferred that tree-fallowing commenced around 1200 cal BP based on increased frequencies of *Casuarina* pollen. Subsequent palaeoecological research at Ambra Crater noted increases in woody disturbance taxa, *Casuarina*, fern spores and Cyperaceae from 1690/1420 cal BP; these were interpreted to represent increased intensity of human settlement, potentially including the beginning of tree-fallowing (Sniderman et al. 2009: 456). Archaeologically, *Casuarina* wood has been identified sporadically in different stratigraphic layers at Kuk (Denham 2003b; cf. Powell 1982b), but is present in much greater numbers in the fills of ditches post-dating the fall of Olgaboli tephra at *c.* 1200–1000 cal BP (Coulter et al. 2009).

Tree-fallowing, like the planting of legume rotations such as winged bean (*Psophocarpus tetragonolobus*), is a technique designed to restore soil fertility, thereby increasing productivity on valley slopes. The impacts of these techniques, however, would be relatively mild and are unlikely to have initiated an abandonment of wetland cultivation. Even though it seems to have been invented at least 1000 years ago, *Casuarina* tree-fallowing is only practised in a relatively limited geographical area, having spread slowly across the highlands (Bourke & Harwood 2009).

Intensive Sweet Potato Cultivation (*Ipomoea Batatas*)
and Pig Husbandry (*Sus* sp.) (Adoption)

Pigs (*Sus* sp.) are intimately associated with ethnographic portrayals of highlander societies. Pigs were a fundamental aspect of highlander social life; they were a means through which highlanders could store surplus and may have formed the basis for the emergence of big-men societies (Modjeska 1982). Big-men were able to accumulate prestige through control of the circulation of pigs and other valuables in traditional exchange ceremonies (Strathern 1971).

The domesticated pigs that are found on New Guinea belong to a clade originally domesticated in mainland Asia (Larson et al. 2007). Despite the significance of pigs to ethnographic portrayals of highlander societies, it is unlikely that pigs were present, or present in any number, in the highlands before approximately 2000 years ago. In a recent review, Sutton et al. (2009: 53) state "archaeozoological evidence is inadequate to address questions of when pigs and dogs were introduced to New Guinea, although both were clearly present by 1500 years ago."

Pigs may only have become important to highlander social and economic life following the introduction of another exotic, the South American domesticate sweet potato (*Ipomoea batatas*). Sweet potato is well adapted to grow in the highlands because it outperforms taro in poorer soils at altitude (Bourke & Harwood 2009). Sweet potato is a relatively ubiquitous staple in the main intermontane valleys and is considered to have replaced taro (*Colocasia esculenta*) as the principal crop. Sweet potato was a Polynesian introduction to the Pacific, sometime after 1000 cal BP, although it is generally considered to be a post-Magellan introduction to New Guinea (post-AD 1500s; Golson 1977).

Even more intriguing than seeking to date the introduction of these two exotics is to consider the speed at which they were incorporated into highlander socioeconomic life. The integration of sweet potato into existing systems of cultivation seems relatively straightforward and probably occurred rapidly: it is a vegetatively propagated tuberous plant like taro and yams. Conversely, highlanders, like the rest of New Guinea's inhabitants, had not previously domesticated any indigenous animals. Their only experiences with domesticated animals are the dog (*Canis familiaris*) and chicken (*Gallus gallus*), both of Asian derivation and both potentially introduced around the same time as pigs in the Late Holocene (Sutton et al. 2009). The adoption of pigs would have required a major shift in thinking and land use. Rather than hunting animals, or possibly capturing them for later trade and consumption, e.g. cassowaries (*Casuarius* spp.), people would have kept, bred and raised pigs. Pigs need to be fed, fenced and protected. The adoption of pig husbandry by highlanders was probably slow initially; their number and social significance probably expanded with the widespread adoption of sweet potato within the last 500 years.

The sweet potato was only one of many post-Magellan introductions to the highlands (and presumably there were many more pre-Magellan introductions throughout the Holocene that have not yet been recorded). Ideas and things moved along local-scale social networks that cumulatively resulted in their transfer from the coast to the highlands. Although these social networks seem to have existed since the Pleistocene, they became more intensive and extensive during the Late Holocene (Hughes 1977). For example, marine shells have been traded into the interior of the island for thousands of years and were a major source of wealth and status in highland societies by the mid-twentieth century. Numerous other introductions or forms of indirect contact occurred ahead of direct contact between highlanders and non–New Guineans, including metal blades, tobacco (*Nicotania tobacum*), food crops and various social practices (Bulmer 1967, 1982; Majnep & Bulmer 1977).

WHY ADOPT?

The above chronologies of practice and forms of plant exploitation have focused on documenting the long-term history of innovations and introductions. There is very little discussion of why people adopted new technologies and practices. Ultimate causation is extremely difficult to elicit in historical research; there are always alternative interpretations. Furthermore, it is unlikely that a single cause was important at all times in the past; rather, different reasons for adopting a particular practice will have influenced different people in specific historic-geographic contexts.

For example, the increased drainage of wetlands over the last 4000 years could be attributed to climatic stresses accompanying intensification of ENSO during the Late Holocene. In the highlands, El Niño years give rise to droughts, which have caused major food shortages in recent decades (Allen 2000). At these times wetland plots are centres of agricultural productivity because of the collapse in dryland production. People maintain wetland plots, in part, as a risk reduction strategy (see Ballard 1995). Even though the increasingly widespread adoption of wetland drainage technology may be linked to climate, this does not necessarily account for its inception.

Any attempt to elicit rationales for innovations, introductions and adoptions in horticultural practices in the Upper Wahgi Valley requires a consideration of multiple, intersecting factors. Environmental phenomena include climatic fluctuations and at least six volcanic ash fall events (Coulter et al. 2009). Social phenomena include demography, transformations in status and political structures, encounters with other groups, and new ideas and ways of doing things. Other factors lie at the nexus of human–environmental relationships, such as landscape degradation, which includes forest clearance and disturbance, as well as soil depletion. Although patterns may be identified in the timing of particular events, correlations do not necessarily give rise to causation (Gould 1970).

FROM PAST TO FUTURE

The practice-centred method advanced here can be applied to horticultural innovations, introductions and adoptions in the present as readily as to those in the past. It provides a methodological framework that can encapsulate agronomic and ethnographic information from the recent past, as well as multidisciplinary data from the long-term histories of archaeology, geomorphology and palaeoecology. In doing so, it provides a means to make information from the past relevant to the present and future.

An understanding of how people reacted to environmental and social challenges in the past has a bearing on understanding how traditional horticultural practices can be managed today. Such analyses can be expanded to track or develop hypotheses regarding crop domestication in the past that are relevant for breeding programs seeking to improve cultivars for future generations. There is also an inherent historical value in studying the development of horticulture in the highlands: it gives Papua New Guineans a collective sense of self and identity in their rapidly changing world.

REFERENCES

Allen, B. 2000. The 1997–98 Papua New Guinea drought: perceptions of disaster, in R.H. Grove & J. Chappell (ed.) *El Niño—history and crisis*: 109–22. Cambridge: White Horse Press.

Arroyo-Kalin, M. 2008. Steps towards an ecology of landscape: the pedo-stratigraphy of anthropogenic dark earths, in W. Woods, W. Teixeira, J. Lehmann, C. Steiner, A. WinklerPrins & L. Rebellato (ed.) *Amazonian dark earths: Wim Sombroek's vision*: 33–83. Dordrecht: Kluwer.

Balée, W. 1994. *Footprints of the forest: Ka'apor ethnobotany*. New York: Columbia University Press.

Ballard, C. 1995. The death of a great land: ritual, history and subsistence revolution in the Southern Highlands of Papua New Guinea. Unpublished PhD dissertation, Australian National University.

Ballard, C. 2001. Wetland drainage and agricultural transformations in the Southern Highlands of Papua New Guinea. *Asia Pacific Viewpoint* 42: 287–304.

Bayliss-Smith, T.P. 2007. The meaning of ditches: interpreting the archaeological record using insights from ethnography, in T.P. Denham, J. Iriarte & L. Vrydaghs (ed.) *Rethinking agriculture: archaeological and ethnoarchaeological perspectives*: 126–48. Walnut Creek: Left Coast Press.

Bourdieu, P. 1990. *The logic of practice*. Cambridge: Polity Press.

Bourke, R.M. 2001. Intensification of agricultural systems in Papua New Guinea. *Asia Pacific Viewpoint* 42: 219–36.

Bourke, R.M. & T. Harwood (ed.). 2009. *Food and agriculture in Papua New Guinea*. Canberra: ANU E Press.

Brookfield, H.C. & P. Brown. 1963. *Struggle for land*. Melbourne: Oxford University Press.

Bulmer, R.N.H. 1967. Why is the cassowary not a bird? A problem of zoological taxonomy among the Karam of the New Guinea highlands. *Man* (NS) 2: 5–25.

Bulmer, R.N.H. 1982. Crop introductions and their consequences in the Upper Kaironk Valley, Simbai area, Madang Province, in R.M. Bourke & V. Kesavan (ed.) *Proceedings of the second Papua New Guinea food crops conference*, volume 2: 282–88. Port Moresby: DPI.

Bulmer, R.N.H. 2005. Reflections in stone: axes and the beginnings of agriculture in the Central Highlands of New Guinea, in A. Pawley, R. Attenborough, J. Golson & R. Hide (ed.) *Papuan pasts: cultural, linguistic and biological histories of Papuan-speaking peoples*: 387–450. Canberra: Australian National University.

Caballero, J. 2004. Patterns in human–plant interaction: an evolutionary perspective. Paper presented at the International Society of Ethnobiology, Ninth International Congress, Canterbury, 13–17 June 2004.

Christensen, O.A. 1975. Hunters and horticulturalists: a preliminary report of the 1972–4 excavations in the Manim Valley, Papua New Guinea. *Mankind* 10: 24–36.

Clarke, W.C. 1971. *Place and people: an ecology of a New Guinea community.* Berkeley: University of California Press.

Coulter, S., T.P. Denham, C. Turney & V. Hall. 2009. The geochemical characterisation and correlation of locally distributed Late Holocene tephra layers at Ambra Crater and Kuk Swamp, Papua New Guinea. *Geological Journal* 44: 568–92.

Denevan, W.M. 2001. *Cultivated landscapes of native Amazonia and the Andes.* Oxford: Oxford University Press.

Denham, T.P. 2003a. Archaeological evidence for mid-Holocene agriculture in the interior of Papua New Guinea: a critical review. *Archaeology in Oceania* (Special Issue) 38: 159–76.

Denham, T.P. 2003b. The Kuk morass: multi-disciplinary investigations of early- to mid-Holocene plant exploitation at Kuk Swamp, Wahgi Valley, Papua New Guinea. Unpublished PhD dissertation, Australian National University.

Denham, T.P. 2005a. Agricultural origins and the emergence of rectilinear ditch networks in the highlands of New Guinea, in A. Pawley, R. Attenborough, J. Golson & R. Hide (ed.) *Papuan pasts: cultural, linguistic and biological histories of Papuan-speaking peoples*: 329–61. Canberra: Australian National University.

Denham, T.P. 2005b. Envisaging early agriculture in the highlands of New Guinea: landscapes, plants and practices. *World Archaeology* 37: 290–306.

Denham, T.P. 2007a. Thinking about plant exploitation in New Guinea: towards a contingent interpretation of agriculture, in T.P. Denham, J. Iriarte & L. Vrydaghs (ed.) *Rethinking agriculture: archaeological and ethnoarchaeological perspectives*: 78–108. Walnut Creek: Left Coast Press.

Denham, T.P. 2007b. Exploiting diversity: plant exploitation and occupation in the interior of New Guinea during the Pleistocene. *Archaeology in Oceania* 42: 41–48.

Denham, T.P. 2008a. Environmental archaeology: interpreting practices-in-the-landscape through geoarchaeology, in B. David & J. Thomas (ed.) *Handbook of landscape archaeology*: 468–81. Walnut Creek: Left Coast Press.

Denham, T.P. 2008b. Traditional forms of plant exploitation in Australia and New Guinea: the search for common ground. *Vegetation History and Archaeobotany* 17: 245–48.

Denham, T.P. 2009. A practice-centred method for charting the emergence and transformation of agriculture. *Current Anthropology* 50: 661–67.

Denham, T.P. 2011. Early agriculture and plant domestication in New Guinea and island Southeast Asia. *Current Anthropology* 52(S4): S379–95.

Denham, T.P. & H. Barton. 2006. The emergence of agriculture in New Guinea: continuity from pre-existing foraging practices, in D.J. Kennett & B. Winterhalder (ed.) *Behavioral ecology and the transition to agriculture*: 237–64. Berkeley: University of California Press.

Denham, T.P., R. Fullagar & L. Head. 2009a. Plant exploitation on Sahul: from colonisation to the emergence of regional specialisation during the Holocene. *Quaternary International* 202: 29–40.

Denham, T.P., J. Golson & P.J. Hughes. 2004a. Reading early agriculture at Kuk (Phases 1–3), Wahgi Valley, Papua New Guinea: the wetland archaeological features. *Proceedings of the Prehistoric Society* 70: 259–98.

Denham, T.P. & S.G. Haberle. 2008. Agricultural emergence and transformation in the Upper Wahgi Valley during the Holocene: theory, method and practice. *The Holocene* 18: 499–514.

Denham, T.P., S.G. Haberle & C. Lentfer. 2004b. New evidence and interpretations for early agriculture in highland New Guinea. *Antiquity* 78: 839–57.

Denham, T.P., S.G. Haberle, C. Lentfer, R. Fullagar, J. Field, M. Therin, N. Porch & B. Winsborough. 2003. Origins of agriculture at Kuk Swamp in the highlands of New Guinea. *Science* 301: 189–93.

Denham, T.P., J. Iriarte & L. Vrydaghs (ed.). 2007. *Rethinking agriculture: archaeological and ethnoarchaeological perspectives.* Walnut Creek: Left Coast Press.

Denham, T.P., K. Sniderman, K. Saunders, B. Winsborough & A. Pierret. 2009b. Contiguous multi-proxy analyses (X-radiography, diatom, pollen and microcharcoal) of Holocene archaeological features at Kuk Swamp, Upper Wahgi Valley, Papua New Guinea. *Geoarchaeology* 24: 715–42.

Eco, U. 1991. *Interpretation and overinterpretation: world, history, texts.* Cambridge: Cambridge University Press.

Fairbairn, A. 2005. An archaeobotanical perspective on Holocene plant use practices in lowland northern New Guinea. *World Archaeology* 37: 487–502.

Fullagar, R., J. Field, T.P. Denham & C. Lentfer. 2006. Early and mid-Holocene processing of taro (*Colocasia esculenta*) and yam (*Dioscorea* sp.) at Kuk Swamp in the highlands of Papua New Guinea. *Journal of Archaeological Science* 33: 595–614.

Gepts P. & R. Papa. 2002. Evolution during domestication, in *Encyclopedia of life sciences.* London: John Wiley. DOI: 10.1038/npg.els.0003071.

Golson, J. 1977. No room at the top: Agricultural intensification in the New Guinea highlands, in J. Allen, J. Golson & R. Jones (ed.) *Sunda and Sahul: prehistoric studies in Southeast Asia, Melanesia and Australia*: 601–38. London: Academic.

Golson, J. 1982. The Ipomoean revolution revisited: society and sweet potato in the upper Wahgi Valley, in A. Strathern (ed.) *Inequality in New Guinea highland societies*: 109–36. Cambridge: Cambridge University Press.

Golson, J. 1991. The New Guinea highlands on the eve of agriculture. *Bulletin of the Indo-Pacific Prehistory Association* 11: 82–91.

Golson, J. 1997. The Tambul spade, in H. Levine & A. Ploeg (ed.) *Work in progress: essays in New Guinea highlands ethnography in honour of Paula Brown Glick*: 142–71. Oxford: Peter Lang.

Golson, J. 2002. Gourds in New Guinea, Asia and the Pacific, in S. Bedford, C. Sand & D. Burley (ed.) *Fifty years in the field. Essays in honour and celebration of Richard Shutler Jr.'s archaeological career*: 69–78. (New Zealand Archaeological Journal Monograph 25). Auckland: Auckland Museum.

Golson, J. & A. Steensberg. 1985. The tools of agricultural intensification in the New Guinea highlands, in I. Farrington (ed.) *Prehistoric intensive agriculture in the tropics*: 247–84 (British Archaeological Reports International Series 232). Oxford: British Archaeological Reports.

Golson, J., R.J. Lampert, J.M. Wheeler & W.R. Ambrose. 1967. A note on carbon dates for horticulture in the New Guinea highlands. *Journal of the Polynesian Society* 76: 369–71.

Gorecki, P. 1979. Population growth and abandonment of swamplands: a New Guinea highlands example. *Journal de la Société des Océanistes* 35: 97–107.

Gott, B. 2005. Aboriginal fire management in south-eastern Australia: aims and frequency. *Journal of Biogeography* 32: 1203–8.

Gould, P. 1970. Is *statistix inferens* the geographical name for a wild goose? *Economic Geography* 46: 439–48.

Gregory, D.J. 2000. Time geography, in R.J. Johnston, D.J. Gregory, G. Pratt & M.J. Watts (ed.) *The dictionary of human geography*: 830–33. Oxford: Blackwell.

Gremillion, K.J. & D.R. Piperno 2009. Human behavioural ecology, phenotypic (developmental) plasticity and agricultural origins. *Current Anthropology* 50: 615–19.

Groube, L. 1989. The taming of the rainforests: a model for Late Pleistocene forest exploitation in New Guinea, in D.R. Harris & G.C. Hillman (ed.) *Foraging and farming: the evolution of plant exploitation*: 292–304. London: Unwin Hyman.

Groube, L., J. Chappell, J. Muke & D. Price. 1986. A 40,000 year-old human occupation site at Huon Peninsula, Papua New Guinea. *Nature* 324: 453–35.
Haberle, S.G. 1994. Anthropogenic indicators in pollen diagrams: problems and prospects for late Quaternary palynology in New Guinea, in J.G. Hather (ed.) *Tropical archaeobotany: applications and new developments*: 172–201. London: Routledge.
Haberle, S.G. 2003. The emergence of an agricultural landscape in the highlands of New Guinea. *Archaeology in Oceania* 38: 149–58.
Haberle, S.G. 2007. Prehistoric human impact on rainforest biodiversity in highland New Guinea. *Philosophical Transactions of the Royal Society (B)* 362: 219–28.
Haberle, S.G., C. Lentfer, S. O'Donnell & T.P. Denham. 2012. The palaeoenvironments of Kuk Swamp from the beginnings of agriculture in the highlands of Papua New Guinea. *Quaternary International* 249: 129–39.
Hägerstrand, T. 1970. What about people in regional science? *Papers of the Regional Science Association* 24: 7–21.
Harris, D.R. 1989. An evolutionary continuum of people–plant interaction, in D.R. Harris & G.C. Hillman (ed.) *Foraging and farming: the evolution of plant exploitation*: 11–26. London: Unwin Hyman
Harris, D.R. 1996. *The origins and spread of agriculture and pastoralism in Eurasia*. London: University College London Press.
Harris, D.R. & P.J. Hughes. 1978. An early agricultural system at Mugumamp Ridge, Western Highlands Province, Papua New Guinea. *Mankind* 11: 437–45.
Hitchcock, G. 2010. Mound-and-ditch taro gardens of the Bensbach or Torassi River area, southwest Papua New Guinea. *The Artefact* 33: 70–90.
Hope, G.S. 2009. Environmental change and fire in the Owen Stanley Ranges, Papua New Guinea. *Quaternary Science Reviews* 28: 2261–76.
Hope, G.S. & S.G. Haberle. 2005. The history of the human landscapes of New Guinea, in A. Pawley, R. Attenborough, J. Golson & R. Hide (ed.) *Papuan pasts: cultural, linguistic and biological histories of Papuan-speaking peoples*: 541–54 (Pacific Linguistics 572). Canberra: Australian National University.
Hughes, I. 1977. *New Guinea Stone Age trade* (Terra Australis 3). Canberra: Australian National University.
Hughes, P.J., M.E. Sullivan & D. Yok. 1991. Human induced erosion in a highlands catchment in Papua New Guinea: the prehistoric and contemporary records. *Zeitschrift für Geomorphologie Suppl.* 83: 227–39.
Hughes, P.J., T.P. Denham & J. Golson. In press. The Kuk Swamp, in J. Golson, T.P. Denham, P. Swadling & J. Muke (ed.) *10,000 years of gardening*. Canberra: ANU E Press.
Jones, M.K. & T. Brown. 2007. Selection, cultivation and reproductive isolation: a reconsideration of the morphological and molecular signals of domestication, in T.P. Denham, J. Iriarte & L. Vrydaghs (ed.) *Rethinking agriculture: archaeological and ethnoarchaeological perspectives*: 36–49. Walnut Creek: Left Coast Press.
Kennedy, J. & W. Clarke. 2004. *Cultivated landscapes of the southwest Pacific* (RMAP Working Paper 50). Canberra: Research School of Pacific and Asian Studies, Australian National University.
Ketan, J. 1998. *An ethnohistory of Kuk*. Port Moresby: NRI.
Larson, G., T. Cucchi, M. Fujita, E. Matisoo-Smith, J. Robins, A. Anderson, B. Rolett, M. Spriggs, G. Dolman, T-H. Kim, N. Thi Dieu Thuy, E. Randi, M. Doherty, R. Awe Due, R. Bollt, T. Djubiantono, B. Griffin, M. Intoh, E. Keane, P.V. Kirch, K-T. Li, M. Morwood, L.M. Pedriña, P.J. Piper, R.J. Rabett, P. Shooter, G. Van Den Bergh, E. West, S. Wickler, J. Yuan, A. Cooper & K. Dobney. 2007. Phylogeny and ancient DNA of Sus provides insights into Neolithic expansion in Island Southeast Asia and Oceania. *Proceedings of the National Academy of Sciences (Washington DC)* 104: 4834–39.

Latinis, D.K. 2000. The development of subsistence system models for Island Southeast Asia and Near Oceania: the nature and role of arboriculture and arboreal-based economies. *World Archaeology* 32: 41–67.

Leahy, M. 1936. The Central Highlands of New Guinea. *Geographical Journal* 87: 229–62.

Lebot, V. 1999. Biomolecular evidence for plant domestication in Sahul. *Genetic Resources and Crop Evolution* 46: 619–28.

Majnep, I.S. & R.N.H. Bulmer. 1977. *Birds of my Kalam country*. Auckland: Auckland University Press.

Marshall, F. 2007. African pastoral perspectives on domestication of the donkey, in T.P. Denham, J. Iriarte & L. Vrydaghs (ed.) *Rethinking agriculture: archaeological and ethnoarchaeological perspectives*: 371–407. Walnut Creek: Left Coast Press.

Modjeska, N. 1982. Production and inequality: perspectives from central New Guinea, in A. Strathern (ed.) *Inequality in New Guinea highland societies*: 50–108. Cambridge: Cambridge University Press.

Mountain, M-J. 1991. Highland New Guinea hunter–gatherers: the evidence of Nombe Rockshelter, Simbu, with emphasis on the Pleistocene. Unpublished PhD dissertation, Australian National University.

Moy, C.M., G.O. Seltzer, D. Rodbell & D.M. Anderson. 2002. Variability of El Niño/Southern Oscillation activity at millennial timescales during the Holocene epoch. *Nature* 420: 162–65.

Muke, J. & H. Mandui. 2003. In the shadows of Kuk: evidence for prehistoric agriculture at Kana, Wahgi Valley, Papua New Guinea. *Archaeology in Oceania* 38: 177–85.

Pearsall, D.M. 2007. Modelling prehistoric agriculture through the palaeoenvironmental record: theoretical and methodological issues, in T.P. Denham, J. Iriarte & L. Vrydaghs (ed.) *Rethinking agriculture: archaeological and ethnoarchaeological perspectives*: 210–30. Walnut Creek: Left Coast Press.

Pétrequin, A-M. & P. Pétrequin. 2006. *Objets de pouvoir en Nouvelle Guinée*. Paris: Réunion des Musées Nationaux.

Piperno, D.R. & D.M. Pearsall. 1998. *The origins of agriculture in the lowland neotropics*. San Diego: Academic Press.

Powell, J.M. 1982a. The history of plant use and man's impact on the vegetation, in J.L. Gressitt (ed.) *Biogeography and ecology of New Guinea*, volume I: 207–27. The Hague: Junk.

Powell, J.M. 1982b. Plant resources and palaeobotanical evidence for plant use in the Papua New Guinea highlands. *Archaeology in Oceania* 17: 28–37.

Powell, J.M., A. Kulunga, R. Moge, C. Pono, F. Zimike & J. Golson. 1975. *Agricultural traditions in the Mount Hagen area* (Department of Geography Occasional Paper 12). Port Moresby: University of Papua New Guinea.

Roscoe, P. 2002. The hunters and gatherers of New Guinea. *Current Anthropology* 43: 153–62.

Serpenti, L.M. 1965. *Cultivators in the swamps*. Assen: Van Gorcum.

Sillitoe, P. 1989. *Made in Niugini: technology in the highlands of Papua New Guinea*. London: British Museum.

Smith, B.D. 2001. Low-level food production. *Journal of Archaeological Research* 9: 1–43.

Sniderman, J.M.K., J. Finn & T.P. Denham. 2009. A late Holocene palaeoecological record from Ambra Crater in the highlands of Papua New Guinea and implications for agricultural history. *The Holocene* 19: 449–58.

Spriggs, M. 1996. Early agriculture and what went before in Island Melanesia: continuity or intrusion?, in D.R. Harris (ed.) *The origins and spread of agriculture and pastoralism in Eurasia*: 524–37. London: University College London Press.

Steensberg, A. 1980. *New Guinea gardens: a study of husbandry with parallels in prehistoric Europe*. London: Academic Press.

Strathern, A. 1971. *The rope of Moka*. Cambridge: Cambridge University Press.

Summerhayes, G.R., M. Leavesley, A. Fairbairn, H. Mandui, J. Field, A. Ford & R. Fullagar. 2010. Human adaptation and plant use in highland New Guinea 49,000 to 44,000 years ago. *Science* 330: 78–81.

Sutton, A., M-J. Mountain, K. Aplin, S. Bulmer & T.P. Denham. 2009. Archaeozoological records for the highlands of New Guinea: a review of current evidence. *Australian Archaeology* 69: 41–58.

Terrell, J.E. 2002. Tropical agroforestry, coastal lagoons and Holocene prehistory in Greater Near Oceania, in Y. Shuji & P.J. Matthews (ed.) *Proceedings of the International Area Studies Conference VII: vegeculture in eastern Asia and Oceania*: 195–216. Osaka: National Museum of Ethnology.

Terrell, J.E., J.P. Hart, S. Barut, N. Cellinese, A. Curet, T.P. Denham, H. Haines, C.M. Kusimba, K. Latinis, R. Oka, J. Palka, M.E.D. Pohl, K.O. Pope, J.E. Staller & P.R. Williams. 2003. Domesticated landscapes: the subsistence ecology of plant and animal domestication. *Journal of Archaeological Method and Theory* 10: 323–68.

Waddell, E. 1972. *The mound builders: agricultural practices, environment and society in the Central Highlands of New Guinea*. Seattle and London: University of Washington Press.

Weiner, J.F. 1991. *The empty place*. Bloomington: Indiana University Press.

Wylie, A. 1985. The reaction against analogy, in M.B. Schiffer (ed.) *Advances in Archaeological Method and Theory* 8: 63–111. New York: Academic Press.

Yen, D.E. 1989. The domestication of environment, in D.R. Harris & G.C. Hillman (ed.) *Foraging and farming: the evolution of plant exploitation*: 55–75. London: Unwin Hyman.

Yen, D.E. 1991. Domestication: the lessons from New Guinea, in A. Pawley (ed.) *Man and a half: essays in Pacific anthropology and ethnobiology in honour of Ralph Bulmer*: 558–69. Auckland: The Polynesian Society.

Yen, D.E. 1995. The development of Sahul agriculture with Australia as bystander. *Antiquity* 69: 831–47.

Yen, D.E. 1996. Melanesian arboriculture: historical perspectives with emphasis on the genus *Canarium*, in B.R. Evans, R.M. Bourke & P. Ferrar (ed.) *South Pacific indigenous nuts*: 36–44. Canberra: ACIAR.

8 Agricultural Economies and Pyrotechnologies in Bronze Age Jordan and Cyprus

Steven E. Falconer and Patricia L. Fall

INTRODUCTION

The development of early civilisations in the eastern Mediterranean and Near East is particularly noteworthy for the variety of paths whereby agrarian societies became increasingly differentiated, often involving the periodic amalgamation and abandonment of urban communities. Following a deeply rooted intellectual tradition (e.g. Childe 1950; Smith 2009), scholars have long envisioned cities as the nuclei that integrated central places with each other and with the myriad villages that housed the majority of ancient populations. A comparison of Bronze Age communities in the Jordan Rift and on the island of Cyprus provides a perspective on emerging complex societies from an alternative vantage point focused on the interactions between farming communities, their managed environments and pyrotechnologies in distinctly non-urban social settings.

The late prehistory of both the southern Levant and Cyprus also fit well, but intriguingly differently, within the interpretive paradigm of the secondary products revolution (Sherratt 1981; Greenfield 2010). The Levantine Bronze Age featured the advent of towns and exchange of secondary products in Early Bronze II–III, their abandonment during Early Bronze IV and a dramatic rejuvenation of cities in the Middle Bronze Age (see Levy 1994). Over a roughly comparable time span, ancient Cypriot society witnessed intensified production of secondary goods through the Early and Middle Cypriot periods (e.g. Knapp 2008: 74–82; see also Webb, this volume), and towards the end of the Bronze Age the emergence of Late Cypriot urban markets and international exchange (Knapp 2008: 66–74, 131–59). Early/ Middle Cypriot domestic evidence stems primarily from relatively few previously excavated communities, most notably Marki-*Alonia* (Frankel & Webb 1996, 2006), Sotira-*Kaminoudhia* (Swiny et al. 2003) and Alambra-*Mouttes* (Coleman et al. 1996). Our study takes advantage of these social and economic dynamics to compare roughly contemporaneous agrarian communities in distinctly different social, economic and environmental contexts (Table 8.1). Politiko-*Troullia* illustrates pre-urban life and landscape on Cyprus (see Webb 2009 and Frankel & Webb 2012a on households at

Table 8.1   Bronze Age chronologies for the southern Levant and Cyprus, showing general chronological relationships between Tell el-Hayyat, Tell Abu en-Ni'aj and Politiko-*Troullia*

Southern Levant	Date	Cyprus
Late Bronze urbanism		Late Cypriot urbanisation
	c. 1500 BC	
Middle Bronze re-urbanisation (Tell el-Hayyat)		Middle Cypriot pre-urban (Politiko-*Troullia*)
	c. 2000 BC	
Early Bronze IV urban collapse (Tell Abu en-Ni'aj)		Early Cypriot pre-urban (Politiko-*Troullia*)
	c. 2400 BC	
Early Bronze II–III urbanisation		Chalcolithic pre-urban
	c. 3000 BC	

Marki-*Alonia*), while Tell Abu en-Ni'aj and Tell el-Hayyat exemplify agrarian ruralism in the southern Levant during urban abandonment and redevelopment, respectively.

BRONZE AGE COMMUNITIES IN THE JORDAN RIFT AND CYPRUS

Tell Abu en-Ni'aj and Tell el-Hayyat embody the remains of Bronze Age farming villages in the *ghor*, the fertile agricultural lowlands just above the active stream channel of the Jordan River (Figure 8.1). These low mounds (3.5–4.5m high) lie only 1.5km apart at approximately 250m below sea level (Ibrahim et al. 1976: 51–4, sites 56 and 64). If their entire areas were occupied, population densities in traditional Middle Eastern villages (generally 200–250 people/ha, see e.g. Kramer 1982) suggest that Ni'aj (2.5ha) had 500 to 600 inhabitants, while Hayyat (0.5ha) housed 100 to 125 people. Tell Abu en-Ni'aj documents agrarian life during the abandonment of Levantine urbanism in Early Bronze IV (Falconer & Fall 2009). Excavated evidence stems from seven stratified phases of mud-brick houses and sherd-paved streets in the heart of the village and an apparent pressing installation on its eastern flank. Tell el-Hayyat illustrates village life amid re-urbanised society through the full course of the Middle Bronze Age (Falconer & Fall 2006). Hayyat features six phases of stratified stone and mud-brick architecture, including the sequential remains of four Canaanite

Figure 8.1 Map locating the Bronze Age settlements of Tell el-Hayyat and Tell Abu en-Ni'aj, Jordan, and Politiko-*Troullia*, Cyprus

temples and a well-preserved updraft pottery kiln built into the tell's southern slope.

Politiko-*Troullia* lies at the interface between the fertile Mesaoria Plain of central Cyprus and the copper-bearing Troodos Mountain foothills. Concentrated Bronze Age surface ceramics, plus clear signatures of buried architecture revealed by soil resistivity, indicate a settlement covering at least 2ha (Falconer et al. 2005). More than 3m of stratified remains document village life in the prehistoric Bronze Age (beginning late in the Early Cypriot period and continuing into the Middle Cypriot period; Falconer & Fall in press). Excavations in Politiko-*Troullia* East unearthed a compound of domestic structures and outbuildings associated with an exterior metallurgical workshop (Fall et al. 2008). Politiko-*Troullia* West, situated roughly 150m across the site, revealed two adjacent courtyards bounded on the south by a walled alleyway and its deeply stratified refuse-laden surfaces, and on the west, north and east by adjoining rooms. Spatial analysis of agricultural terracing and associated material culture on adjacent hillsides (covering approximately 20ha) infers long-term management of the surrounding landscape, most intensively during the prehistoric Bronze Age and Iron Age, with an apparent hiatus during the Late Cypriot advent of urbanism (Fall et al. 2012).

AVENUES OF INFERENCE

This study compares patterns of animal exploitation, crop cultivation and pyrotechnological industry as they reveal interactions of society, environment and technology at Politiko-*Troullia*, Tell Abu en-Ni'aj and Tell el-Hayyat. While this evidence varies within sites both temporally and spatially, we concentrate here on overall inter-settlement patterns that provide robust comparisons between these communities in the Levant and Cyprus. The use of animal resources and associated implications for Levantine and Cypriot Bronze Age landscapes may be inferred from the animal bone assemblages recovered from these three settlements. All excavated sediments were screened through 0.5cm mesh screen. The resulting faunal data are presented as NISP (numbers of identified specimens), which provides the best means to compare relative frequencies of animal taxa across time or space (see e.g. O'Connor 2000: 55; Reitz & Wing 2008: 202–10). The bulk of the faunal remains comes from a relatively narrow range of husbanded domesticates: sheep (*Ovis aries*), goat (*Capra hircus*), cattle (*Bos taurus*) and pig (*Sus scrofa*). Wild species are represented at all three sites by minor amounts of wild birds, fish and small mammals, while Mesopotamian fallow deer (*Dama dama mesopotamica*) is found only at Politiko-*Troullia*.

Systematic water flotation of sediment from a variety of excavated contexts at Tell Abu en-Ni'aj, Tell el-Hayyat and Politiko-*Troullia* recovered carbonised botanical remains, including substantial remains of wood charcoal, indicative of crop cultivation, fuel consumption and local vegetation. All sediments showing evidence of charred organic content were sampled and analysed for botanical remains using manual, nonmechanised flotation procedures (see Klinge & Fall 2010). The major plant taxa represented in the floral assemblages from all three sites may be categorised as orchard taxa, cereals, legumes or wild and weedy taxa. Perennial orchard species are mainly olive (*Olea europea*), grape (*Vitus vinifera*) and fig (*Ficus carica*). Cultivated cereals are wheat (*Triticum* sp.), barley (*Hordeum* sp.), oat (*Avena* sp.) and rye (*Secale* sp.). Cultivated legumes include lentil (*Lens culinaris*), garden pea (*Pisum sativum*), chickpea (*Cicer arietinum*), fava bean (*Vicia faba*) and grass pea/bitter vetch (*Lathyrus sativus/Vicia ervilla*). The varieties of wild taxa include wild cereals (Poaceae), wild legumes (Papillionaceae) and *Prosopis*, as well as crop-following weeds (e.g. Asteraceae, *Galium*). These seed remains may be quantified as counts or weights, with seed counts expressed as relative frequencies for each taxon or category. Charcoal remains recovered via flotation are quantified most effectively by weight. Thus density values, reflecting the abundance of carbonised plant remains in flotation samples, are calculated as seed densities (# seeds/kl of processed sediment) and charcoal densities (g charcoal/kl of processed sediment).

Seed and charcoal data, accordingly, may be integrated through the use of seed:charcoal ratios (e.g. based on weights), which permit inference of relative rates of seed vs. charcoal carbonisation. A variety of case studies suggest that ancient societies, particularly those incorporating intensive pyrotechnology (e.g. for metallurgy or firing pottery), tended to exhaust fuel wood sources and subsequently shift to dung fuel (Sillar 2000; Rhode et al. 2007). Ethnographic studies, in turn, reveal that seed carbonisation primarily indicates burning of dung fuel, especially in the domesticated animal economies of the ancient Mediterranean world (e.g. Miller 1996). Seed:charcoal ratios thus reflect varying combinations of fuel wood and dung fuel consumption in antiquity (Miller 1988), with attendant implications for the probable availability of woodland fuel sources.

Bronze Age agrarian society embarked on long-term intensification of previously established technologies. Across the rural landscapes of Cyprus and the Levant, farming communities invested considerable energy in constructing agricultural terraces and check dams to retain field plots and conserve precious soil and rainfall. Their remains demonstrate the ebb and flow of agricultural land use patterns. Within these settlements the use of fuel consumptive pyrotechnologies to produce ceramic containers and metal implements helped create distinctly anthropogenic surrounding landscapes. However, these technologies varied considerably in their material expression and environmental implications. Indeed, variability in all these forms of evidence provides the leitmotif of this study.

ANIMAL MANAGEMENT AND CONSUMPTION

Animal husbandry and consumption at all three settlements focused heavily on domesticated sheep and goat as signalled clearly by bone counts in which *Ovis/Capra* constitute consistently sizable majorities (Table 8.2). The high value for sheep/goat at Tell el-Hayyat reflects, in part, the preference for ovicaprid offerings that were deposited within Hayyat's temple compounds (Falconer & Fall 2006: fig. 6.47). At the other end of the animal exploitation spectrum, the occupants of our two Levantine villages engaged in only negligible hunting and fishing, despite the persistence of the Jordan River as a source of fish and the Jordan Rift as an avian migratory flyway.

Among the remaining taxa, cattle bones found at all three sites indicate a minor component of beef consumption and the role of bovids for the traction needed for plough technology in surrounding fields. The relative frequencies for the remaining taxonomic categories vary widely between settlements, indicating strikingly divergent animal availability and exploitation in the Jordan Rift and Cyprus. At Hayyat and, most strikingly, at Abu en-Ni'aj swine clearly rank second among sources of animal remains. In both settings swine herding provides an attractive complement to sheep/goat herding, since pigs are prolific reproducers, grow quickly to mature meat weight and may be tended as backyard scavengers without the need for extensive pasturing (Zeder 1991: 30; Hesse 1995). On the other hand,

Table 8.2   Relative frequencies of identified bone fragments by animal taxon excavated from Tell Abu en-Ni'aj (Phases 6–1), Tell el-Hayyat (Phases 5–2) and Politiko-*Troullia* (latest phase in *Troullia* East and West). Data from Falconer et al. 2004; Falconer & Fall 2006: tables 5.3 and 5.4; Falconer & Fall in press

Taxon	Tell Abu en-Ni'aj	Tell el-Hayyat	Politiko-*Troullia*
Sheep/goat	60%	71%	68%
Pig	28%	17%	4%
Cattle	11%	12%	8%
Wild	<1%	<1%	20%
NISP	5180	11,970	3863

pigs have relatively high water needs, provide few secondary products and are poorly suited for herding to market (Horwitz & Tchernov 1989; Hesse 1995). Even at Hayyat, in the context of urbanised society, and especially at Ni'aj during urban collapse, pig husbandry appears to be a signature of localised nonmobile animal management in a village environment.

In stark contrast to their Levantine contemporaries, the villagers of Politiko-*Troullia* complemented their sheep/goat herding with ample exploitation of wild animals, most notably through hunting of Mesopotamian fallow deer (*Dama dama mesopotamica*). The importance of deer hunting emerges repeatedly at Early and Middle Cypriot communities elsewhere on Cyprus (Reese 1996; Croft 2003, 2006). This preference is especially pronounced in the generally domestic setting of *Troullia* West, particularly in its alleyway and southern courtyard (Falconer & Fall in press). Thus, while the villagers of the Rift bottomlands augmented the venerable tradition of sheep/goat herding with backyard pig husbandry, the occupants of Politiko-*Troullia* turned to a different supplement by hunting forest-dwelling cervids in the surrounding hill country.

CROP CULTIVATION AND FUEL CONSUMPTION

Relative frequencies of carbonised seeds, as with animal bones, reveal fundamental contrasts between cultivation strategies, especially as practised in the Jordan Rift and on Cyprus (Table 8.3). Cultivated cereals and their attendant crop-following weeds comprise roughly 80% of the relatively similar seed assemblages from Tell Abu en-Ni'aj and Tell el-Hayyat. Orchard taxa from these two sites, despite being preferred for temple offerings at Hayyat (Falconer & Fall 2006: fig. 6.46), contribute much more modestly. In stark contrast, the Politiko-*Troullia* floral profile is dominated overwhelmingly by perennial orchard taxa, with correspondingly much less evidence of wild or weedy taxa. Cereals are represented only modestly, and legumes, which contribute minimally to the array of crops cultivated by the

Table 8.3 Relative frequencies of identified seeds, seed and charcoal densities and seed:charcoal ratios for Tell Abu en-Ni'aj (Phases 6–1), Tell el-Hayyat (Phases 5–2) and Politiko-*Troullia* (latest phase in *Troullia* East and West). Data from Falconer et al. 2004; Falconer & Fall 2006: appendix E; Klinge & Fall 2010; Falconer & Fall in press

Taxon	Tell Abu en-Ni'aj	Tell el-Hayyat	Politiko-Troullia
Orchard	13%	17%	78%
Cereals	33%	30%	6%
Legumes	3%	4%	<1%
Wild/weedy	51%	49%	15%
Identified seeds	2217	8081	401
N samples	60	151	50
Seed density (#/kl)	192,885	16,257	1093
Charcoal density (g/kl)	485	2842	487
Seed:charcoal (g:g)	2.37	0.97	0.35

farmers of all three villages, are virtually absent at Politiko-*Troullia*. While the animal bone assemblages suggest somewhat different modes of herding and hunting at all three sites, the data for carbonised seeds portray similar profiles of crop cultivation around Abu en-Ni'aj and Hayyat while accentuating the differences between this valley-bottom farming and the emphasis on arboriculture in the Troodos foothills around *Troullia*.

Seed:charcoal ratios reveal somewhat more nuanced distinctions in fuel consumption at these three villages, thereby providing a first indication of differing configurations of vegetation availability, fuel consumption and pyrotechnological industry. These ratios produce relatively high values for Tell Abu en-Ni'aj, comparable to those from Bronze Age towns along the Euphrates in Syria (Klinge & Fall 2010). This points to relatively modest burning of fuel wood and heavier use of dung fuel at Abu en-Ni'aj and in the Syrian steppe. The extremely high mean seed density for Abu en-Ni'aj strengthens an interpretation of particularly pervasive reliance on dung fuel by its villagers. When coupled with lower orchard seed frequencies than found at Hayyat or *Troullia*, this evidence suggests Tell Abu en-Ni'aj occupied a largely deforested landscape with minimal availability of fuel wood, whether cut from wild stands or harvested from orchards.

At the other end of the floral spectrum, Politiko-*Troullia* produced the lowest overall seed:charcoal ratio, based on a low mean seed density and a charcoal density comparable to that from Abu en-Ni'aj. However, an abundance of charcoal deposited in and around the courtyard metallurgical workshop at Politiko-*Troullia* elevates the charcoal density for *Troullia* East to 1690g/kl and drops its seed:charcoal ratio to 0.11 (Klinge & Fall

2010), by far the lowest ratio considered in this study. If we incorporate this community's emphasis on arboriculture and the considerable availability of forest-dwelling fallow deer, a snapshot emerges of Politiko-*Troullia* at an ecotone between adjacent managed orchards and presumably nearby woodlands (and mineral resources), amid a very different kind of anthropogenic landscape than that found around Tell Abu en-Ni'aj.

Tell el-Hayyat is situated between Politiko-*Troullia* and Tell Abu en-Ni'aj in terms of mean seed:charcoal ratio and its attendant environmental implications. While Hayyat produced an intermediate seed density, its charcoal density is more than five times higher than those found at the other two settlements. Given its cereal-oriented crop profile, it seems clear that Hayyat's households implemented farming strategies similar to those of Abu en-Ni'aj, probably in a common nonforested local environment given the close proximity of the two sites. However, the charcoal evidence from both Tell el-Hayyat and Politiko-*Troullia* signals the importance of fuel-driven technologies as they impacted their surrounding landscapes. (Preliminary identifications indicate burning of olive wood at Hayyat and pine at *Troullia*.)

PYROTECHNOLOGIES IN THE COUNTRYSIDE

The importance of fuel-consumptive technology for community behaviour is manifested rather differently at Tell Abu en-Ni'aj, Tell el-Hayyat and Politiko-*Troullia*. Although neutron activation analysis suggests that Abu en-Ni'aj and other comparable villages may have produced the signature trickle-painted cups of Early Bronze IV (Falconer 1987), the units excavated at this site revealed no direct evidence of pottery manufacture or use of any other pyrotechnology beyond household fuel use. Indeed, the extremely high mean seed density, low charcoal density and correspondingly high seed:charcoal ratio for Abu en-Ni'aj suggest reliance on dung fuel for domestic needs with little connotation of industrial pyrotechnology.

In contrast to Abu en-Ni'aj, the high charcoal density and very low seed:charcoal ratio for Politiko-*Troullia* East underscore the local availability of firewood, a fuel source favoured for metallurgy over lower-firing animal dung (Forbes 1971; Bamberger & Wincierzt 1990). A roofed courtyard here with charcoal-laden sediments, remnants of a rudimentary furnace, a three-tool carved limestone mould, ceramic crucibles, tap slags and ore fragments suggests household-level industry for the production of utilitarian tools, such as the pins, needles and the dagger hilt found in the alleyway refuse of Politiko-*Troullia* West (Falconer & Fall in press).

The relatively modest seed:charcoal ratio and very high charcoal density (five times that of *Troullia* or Abu en-Ni'aj) signal the importance of pyrotechnology at Tell el-Hayyat, despite its situation in the apparently deforested bottomlands of the Jordan Valley. Here fuel-intensive technology finds multiple expressions: broad-range pottery production and metallurgy involving manufacture of both offertory figurines and utilitarian tools. In addition to the largely intact kiln remains, ceramic wasters and over-fired pots, neutron activation analysis suggests production of cooking pots, storage jars and possibly fine ware bowls, despite Tell el-Hayyat's diminutive size (Falconer 1987). This pattern of pottery distribution stands in contrast to the largely localised production and consumption of pottery in Early and Middle Cypriot communities on Cyprus (Frankel & Webb 2012b).

While Hayyat does not reveal a discernible footprint for a workshop, its metallurgy is clearly multifaceted (Falconer & Fall 2006: 83–95). Although slag and ore fragments are less abundant, probably reflecting greater distances to ore sources than at Politiko-*Troullia*, ceramic crucibles and limestone moulds once produced both everyday tanged implements and anthropomorphic figurines. Interestingly, most manufacturing debris and especially finished metal objects were deposited in the forecourts or interiors of Hayyat's temples, as part of a multimedia array of offertory objects. Thus this element of Bronze Age technology provided a means of interaction not only with the fuel-bearing local environment, but also among the households and social institutions manifested by Hayyat's domestic enclosures and temple compounds.

CONCLUSION

The evidence from Tell el-Hayyat, Tell Abu en-Niʿaj and Politiko-*Troullia* supports a multifaceted portrait of agrarian communities enmeshed in various expressions of the secondary products revolution and in a variety of pre-urban, non-urban or re-urbanised social contexts. Rather than emphasising general similarities among the Bronze Age agrarian societies of Cyprus and the Jordan Rift as they relate to early urbanism (or the lack thereof), this study tends to accentuate the diversity of natural landscapes, plant and animal utilisation, pyrotechnologies and related modes of social interaction in these emerging complex societies. In overview, the villagers of Tell Abu en-Niʿaj practised dedicated sedentary farming on an apparently deforested landscape during a period often interpreted, ironically, in terms of nonsedentary pastoralism. In contrast, the inhabitants of Politiko-*Troullia* utilised a very different anthropogenic landscape of apparently intensive arboriculture in close proximity to nearby forest and mineral resources, which supported relatively expedient household metallurgy. Pyrotechnology plays rather different roles at Tell el-Hayyat, where the importance of utilitarian and especially ritual metallurgy inspired pronounced fuel wood consumption despite its location in an anthropogenic landscape similar to that of Abu en-Niʿaj. Thus, over a roughly common time span in the late third millennium and early second millennium BC, these three sedentary agrarian communities, each populated by no more than a few hundred villagers, illustrate the diversity of social, environmental and technological interactions that characterised the foundations of early civilisation in the eastern Mediterranean.

ACKNOWLEDGMENTS

The excavations and analyses of Tell Abu en-Ni'aj, Tell el-Hayyat and Politiko-*Troullia* were supported by grants from the US National Science Foundation (most recent Award #1031527), US National Endowment for the Humanities, the National Geographic Society and the Wenner-Gren Foundation for Anthropological Research. Research fellowships at the American Center of Oriental Research, Amman (a Council for American Overseas Research Centers Fellowship held by Fall and an ACOR Publication Fellowship held by Falconer), enabled the authors to complete this research. Special thanks are due to the Department of Antiquities, Hashemite Kingdom of Jordan (Directors-General Dr Adnan Hadidi, Dr Ghazi Bisheh and Dr Fawwaz al-Khraysheh), the Department of Antiquities, Republic of Cyprus (Directors Dr Pavlos Flourentzos and Dr Maria Hadjicosti), the American Center of Oriental Research, Amman (Directors Dr David McCreery, Dr Bert DeVries, Dr Pierre Bikai and Dr Barbara Porter), the Cyprus American Archaeological Research Institute, Nicosia (Directors Dr Thomas Davis and Dr Andrew McCarthy), to Mary C. Metzger, zooarchaeologist for the Ni'aj, Hayyat and *Troullia* excavations, and to Wei Ming for drafting our map.

REFERENCES

Bamberger, M. & P. Wincierzt. 1990. Ancient smelting of oxide copper ore, in B. Rothenberg (ed.) *The ancient metallurgy of copper*: 78–122. London: Institute for Archaeo-Metallurgical Studies.

Childe, V.G. 1950. The urban revolution. *Town Planning Review* 21: 3–17.

Coleman, J.E., J.A. Barlow, M.K. Mogelonsky & K.W. Schaar. 1996. *Alambra: a Middle Bronze Age settlement in Cyprus. Archaeological investigations by Cornell University 1974–1985* (Studies in Mediterranean Archaeology 118). Jonsered: Paul Åströms förlag.

Croft, P. 2003. The animal remains, in S. Swiny, G. Rapp & E. Herscher (ed.) *Sotira Kaminoudhia: an Early Bronze Age site in Cyprus* (CAARI Monograph Series 4. ASOR Archaeological Reports 8): 439–48. Boston: American Schools of Oriental Research.

Croft, P. 2006. Animal bones, in D. Frankel & J.M. Webb (ed.) *Marki Alonia. An Early and Middle Bronze Age settlement in Cyprus. Excavations 1995–2000* (Studies in Mediterranean Archaeology 123.2) 263–81. Jonsered: Paul Åströms förlag.

Falconer, S.E. 1987. Village pottery production and exchange: a Jordan Valley perspective, in A. Hadidi (ed.) *Studies in the history and archaeology of Jordan*, volume 3: 251–9. London: Routledge and Kegan Paul.

Falconer, S.E. & P.L. Fall. 2006. *Bronze Age rural economy and village life at Tell el-Hayyat, Jordan* (British Archaeological Reports, International Series 1586). Oxford: Archaeopress.

Falconer, S.E. & P.L. Fall. 2009. Settling the valley: agrarian settlement and interaction along the Jordan Rift during the Bronze Age, in E. Kaptijn & L. Petit (ed.) *A timeless vale: archaeological and related essays on the Jordan Valley*: 97–107 (Archaeological Studies, Leiden University 19). Leiden: Leiden University Press.

Falconer, S.E. & P.L. Fall. In press. Household and Community Behavior at Bronze Age Politiko-*Troullia*, Cyprus. *Journal of Field Archaeology*.

Falconer, S.E., P.L. Fall, T. Davis, M. Horowitz & J. Hunt. 2005. Initial investigations at Politiko *Troullia*, 2004. *Report of the Department of Antiquities, Cyprus* 2005: 69–85.

Falconer, S.E., P.L. Fall, M.C. Metzger & L. Lines. 2004. Bronze Age rural economic transitions in the Jordan Valley. *Annual of the American Schools of Oriental Research* 58: 1–17.

Fall, P.L., S.E. Falconer, C. Galletti, T. Schirmang, E. Ridder & J. Klinge. 2012. Long-term agrarian landscapes in the Troodos foothills, Cyprus. *Journal of Archaeological Science* 39: 2335–47. doi:10.1016/j.jas.2012.02.010

Fall, P.L., S.E. Falconer, M. Horowitz, J. Hunt, M.C. Metzger & D. Ryter. 2008. Bronze Age settlement and landscape of Politiko-*Troullia*, 2005–2007. *Report of the Department of Antiquities, Cyprus* 2008: 183–208.

Forbes, R.J. 1971. *Studies in ancient technology*, volume 3. Leiden: E.J. Brill.

Frankel, D. & J.M. Webb. 1996. *Marki Alonia. An Early and Middle Bronze Age town in Cyprus. Excavations 1990–1994* (Studies in Mediterranean Archaeology 123.1). Jonsered: Paul Åströms förlag.

Frankel, D. & J.M. Webb. 2006. *Marki Alonia. An Early and Middle Bronze Age settlement in Cyprus. Excavations 1995–2000* (Studies in Mediterranean Archaeology 123.2). Jonsered: Paul Åströms förlag.

Frankel, D. & J.M. Webb. 2012a. Household continuity and transformation in a prehistoric Cypriot village, in B.J. Parker & C.P. Foster (ed.) *New perspectives in household archaeology*: 473–500. Winona Lake: Eisenbrauns.

Frankel, D. & J. Webb. 2012b. Pottery production and distribution in prehistoric Bronze Age Cyprus. An application of pXRF analysis. *Journal of Archaeological Science* 39(5): 1380–7. doi:10.1016/j.jas.2011.12.032

Greenfield, H.J. 2010. The secondary products revolution: the past, the present and the future. *World Archaeology* 42: 29–54.

Hesse, B. 1995. Animal husbandry and human diet in the ancient Near East, in J.M. Sasson (ed.) *Civilizations of the ancient Near East*: 203–22. New York: Charles Scribner's Sons.

Horwitz, L. & E. Tchernov. 1989. Animal exploitation in the Early Bronze Age of the southern Levant, in P. de Miroschedji (ed.) *L'urbanisation de la Palestine a l'age du bronze ancien*: 279–96 (British Archaeological Reports, International Series 527). Oxford: Archaeopress.

Ibrahim, M., J.A. Sauer & K. Yassine. 1976. The East Jordan Valley Survey, 1975. *Bulletin of the American Schools of Oriental Research* 222: 41–66.

Klinge, J. & P.L. Fall. 2010. Archaeobotanical inference of Bronze Age land use and land cover in the eastern Mediterranean. *Journal of Archaeological Science* 37: 2622–29.

Knapp, A.B. 2008. *Prehistoric and protohistoric Cyprus: identity, insularity and connectivity*. Oxford: Oxford University Press.

Kramer, C. 1982. *Village ethnoarchaeology: rural Iran in archaeological perspective*. New York: Academic Press.

Levy, T.E. (ed.) 1994. *The archaeology of society in the Holy Land*. New York: Facts on File.

Miller, N.F. 1988. Ratios in paleoethnobotanical analysis, in C.A. Hastorf & V.S. Popper (ed.) *Current paleoethnobotany: analytical methods and cultural interpretations of archaeological plant remains*: 72–85. Chicago: University of Chicago Press.

Miller, N.F. 1996. Seed eaters of the ancient Near East, human or herbivore? *Current Anthropology* 37: 521–8.

O'Connor, T. 2000. *The archaeology of animal bones*. College Station: Texas A&M Press.

Reese, D. 1996. Subsistence economy, in J.E. Coleman, J.A. Barlow, M.K. Mogelonsky & K.W. Schaar (ed.) *Alambra: a Middle Bronze Age settlement in Cyprus. Archaeological investigations by Cornell University 1974–1985*: 217–26 (Studies in Mediterranean Archaeology 118). Jonsered: Paul Åströms förlag.

Reitz, E.J. & E.S. Wing. 2008. *Zooarchaeology* (2nd edition). Cambridge: Cambridge University Press.

Rhode, D., D.B. Madsen, P.J. Brantingham & T. Dargye. 2007. Yaks, yak dung and prehistoric human occupation of the Tibetan Plateau, in D.B. Madsen, C. Fahu & G. Xing (ed.) *Late Quaternary climate change and human adaptation in arid China*: 205–26. Amsterdam: Elsevier.

Sherratt, A.G. 1981. Plough and pastoralism: aspects of the secondary products revolution, in I. Hodder, G. Isaac & N. Hammond (ed.) *Pattern of the past: studies in memory of David Clarke*: 261–305. Cambridge: Cambridge University Press.

Sillar, B. 2000. Dung by preference: the choice of fuel as an example of how Andean pottery production is embedded within wider technical, social, and economic practices. *Archaeometry* 42(1): 43–60.

Smith, M.E. 2009. V. Gordon Childe and the urban revolution: an historical perspective on a revolution in urban studies. *Town Planning Review* 80: 3–29.

Swiny, S., G. Rapp & E. Herscher (ed.). 2003. *Sotira Kaminoudhia: an Early Bronze Age site in Cyprus* (CAARI Monograph Series 4. ASOR Archaeological Reports 8). Boston: American Schools of Oriental Research.

Webb, J.M. 2009. Keeping house: our developing understanding of the Early and Middle Cypriot household (1926–2006). *Medelhavsmuseet. Focus on the Mediterranean* 5: 255–67.

Zeder, M. 1991. *Feeding cities*. Washington, DC: Smithsonian Institution Press.

Changing Technological and Social Environments in the Second Half of the Third Millennium BC in Cyprus

Jennifer M. Webb

INTRODUCTION

Two major archaeologically recognisable cultural entities are visible in mid-third millennium BC Cyprus: an indigenous Late Chalcolithic dependent on hoe-based agriculture and a migrant Philia Early Bronze Age with a radically different social and technological system, including the cattle/plough complex. This was a key point of disjunction in the prehistory of Cyprus, which offered a significant competitive advantage to the newcomers and presented a major adaptive challenge to the pre-existing population. This chapter examines the impact of what appears to have been a relatively sudden introduction of a suite of new technologies and seeks to identify and explain the processes involved in the interaction between Late Chalcolithic and Bronze Age communities and the eventual encompassing of one by the other. It views them as organisationally and ideologically distinct environments—with a focus not so much on the actual physical landscape as on the perceived or experienced environment constituted through previous history and specific cultural tradition and resulting from the constraints and opportunities provided by available technologies and social structures (see Butzer 1972). The incursive Philia system shaped new and divergent sets of material objects and social logics. In an attempt to explain these outcomes, this chapter adopts a contextual approach in order to identify response mechanisms and model the uptake and persistence of technologies and social strategies across the island.

The data examined relate to the latter years of the Late Chalcolithic (2700–2450/2400 BC) and the first phase of the Early Bronze Age (or EBA), known as the Philia Early Cypriot (hereafter EC) (2450/2400–2300/2250 BC). While the absolute chronology is in need of refinement, radiocarbon determinations suggest a significant temporal overlap between these two cultural phases.

THE LATE CHALCOLITHIC/PHILIA EC TRANSITION

The Philia EC was a period of profound social, economic and technological transformation. Prominent amongst the changes visible at this time are innovations in ceramics, architecture and domestic and burial practices and the introduction of cattle, donkeys and possibly new species of goat and sheep as well as plough agriculture and intensive metallurgy. It has long been proposed that these innovations were brought about by a migration to Cyprus from south-western Anatolia or Cilicia (Dikaios 1962: 202–3; Frankel et al. 1995; Webb & Frankel 1999; Frankel 2000; Webb 2002). In a recent paper (Webb & Frankel 2011) it has been specifically identified as a process of colonisation—that is, as a migration involving the planned departure of individuals from one or more communities to establish new communities that replicate those at home (as defined by Manning 2005: 4–7). While use of this term is likely to remain contentious within the Cypriot literature (Knapp 2008: 113), it is clear that the transition to the Bronze Age in Cyprus involved new cultural practices, new technologies and new knowledge and that through processes of encounter, interaction and hybridisation these resulted in the disappearance of an archaeologically identifiable Chalcolithic way of life.

The movement of human groups to Cyprus in the mid-third millennium was clearly based on explicit knowledge of the island, likely acquired during years of prior contact (Peltenburg 2007). It involved a 'transported landscape'—that is, incoming groups brought with them resources that they knew were not available (cattle, donkeys) and targeted those, most notably copper, that they knew were available. Copper probably provided the main incentive, but this need not mean that there was a disruption in access to Anatolian sources. A desire to find new sources of raw materials for the long-distance exchange systems that linked south-eastern Anatolia to the northeast Aegean, the Cyclades and mainland Greece in the early to mid-third millennium (Şahoğlu 2005; Efe 2007), along with advances in maritime technology (Broodbank 2010: 255), may have been critical factors. The point of entry and ongoing authority was almost certainly located on the north coast, probably at Vasilia, a site which has produced a number of elaborate tombs (Hennessy et al. 1988: 25–39). Lead Isotope Analysis suggests that Vasilia was involved in the raw metals trade at this time, both exporting Cypriot copper and importing Cycladic and Anatolian copper and finished artefacts (Webb et al. 2006).

New villages with incursive Philia EC technologies were founded, probably fairly rapidly, in the Ovgos valley, the central lowlands and the foothills of the northern and southern Troodos (Figure 9.1). These villages were located in good agricultural terrain, along natural communication and transport routes and near copper ore bodies, with targeted exploitation of high arsenic ores in the Limassol Forest area perhaps of key importance in explaining movement south of the Troodos range (Georgiou et al. 2011: 359–60). They shared a homogeneous material culture, particularly with regard to ceramic form, decoration and technology, and common sets of metal artefacts and personal ornaments (Bolger 1991: 33–4; Manning 1993: 94;

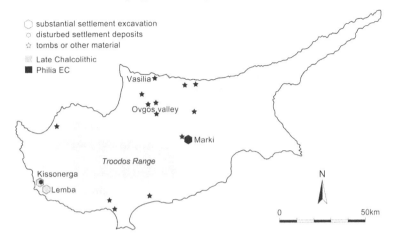

Figure 9.1   Map of Cyprus showing the location of excavated Late Chalcolithic and Philia EC settlement sites and tombs (drawn by the author)

Manning & Swiny 1994: 166; Webb & Frankel 1999; Frankel & Webb 2004). Recent analyses of pottery clays show that most vessels were distributed from the north, perhaps in exchange for copper (Dikomitou 2010, 2011; Dikomitou & Martinòn-Torres 2012). Commensality, and in particular communal drinking, is also indicated by the predominance of pouring and drinking vessels in settlement and mortuary contexts (Webb & Frankel 2008: 289–90, pl. LVIb). These integrative strategies appear to have served as effective mechanisms of alliance and solidarity within and between these dispersed communities—facilitating the procurement and flow of metal, providing access to raw materials and manufactured goods and ensuring the intergenerational transmission of technical and esoteric knowledge.

It should be stated at this point that the database for investigating the Late Chalcolithic/Philia EC transition is limited. Although crucial to understanding mid-third millennium developments, sites with Philia material are still poorly known (Webb & Frankel 1999: 7–13). The only substantially excavated Philia settlement is that of Marki, in the centre of the island (Frankel & Webb 2006). Similarly, Kissonerga and Lemba, in the southwest, are the only two substantially excavated Late Chalcolithic settlements (Peltenburg 1985, 1998). Only Kissonerga provides a view of *in situ* development from the Late Chalcolithic to the Philia EC. The remains of the Philia EC, however, lie within the plough zone and are poorly preserved (Peltenburg 1998: 52–4). Marki, newly founded in the Philia EC, continued in use through the remainder of the EBA. What follows relies heavily on data from these two settlements and in particular Kissonerga.

ADAPTATION, REJECTION AND ENCAPSULATION

What impact did the establishment of a culturally incursive Philia EC settlement network have on existing Chalcolithic communities? Can we identify the processes involved in the meeting and mingling of these technologically and socially diverse groups that led ultimately to the 'disappearance' of the latter from the archaeological record? An increasing dissonance between the real and the perceived environment is certainly likely to have been a critical factor—with the nature and rate of adaptation and/or rejection of new practices by Chalcolithic communities influenced by decisions based on what was known, accessible and valued rather than on any 'objective' reality (Kirch 1980: 112; Knapp & Ashmore 1999; Ashmore 2002; Bender 2002). Viewing the engagement between environment, technology and society in this way requires us to ask how social practices determine which new technologies are absorbed, manipulated or rejected and how these processes are materially expressed and performed (Dobres 2000, 2001).

The Late Chalcolithic and Philia EC 'environments' are, however, for the most part evident only as 'before and after' entities. The migrant Philia EC material culture is one that has already been fully recontextualised within a new physical environment. The earliest stages of this process are likely to have taken place on the north coast and are not yet visible. The Late Chalcolithic, on the contrary, is known almost entirely from the southwest. Processes of adaptation and encapsulation across much of the island are not, therefore, and indeed may never be, observable. We have, as van der Leeuw has noted in relation to the study of change in the archaeological record more generally (1993: 241), a sequence of 'stills'—and interpolating the processes of adaptation, acculturation and hybridisation that took place between them is a significant challenge. Given these inherent difficulties and the small number of excavated sites that sit on either side or cross this transition, a diachronic analysis may offer the most useful perspective on the uptake and persistence of technologies and social strategies across the island. In what follows both chronologically broad and fine-scale data are examined in an attempt to identify pattern and process across this major technological and social divide.

THE DIACHRONIC PERSPECTIVE: IDENTIFYING PATTERN

The more obvious long-term impacts of the arrival of Philia EC communities on Cyprus are not difficult to see. Even the longest-lived Chalcolithic settlements were soon abandoned. At Lemba this was accompanied by site-wide destruction (Peltenburg 1985). The circumstances in which other sites in the south and west were deserted remain unclear. Kissonerga is the only Chalcolithic settlement known to have continued across the Philia transition. The typical Late Chalcolithic pattern of animal exploitation also disappears from view (Figure 9.2a). At Lemba and Kissonerga deer and pig constitute almost 80% of the identifiable bone sample (Croft 1985, 1998). In Philia EC levels at Marki they comprise only 10% while cattle, newly introduced (or re-introduced) to the island by Philia EC communities, are present at 24%, and reliance on caprines, and particularly sheep, is three times higher at 62% (Croft 2006: 263, text table 9.1). A dramatic increase in cereal production is also visible at Marki. Here rubbers and querns (upper and

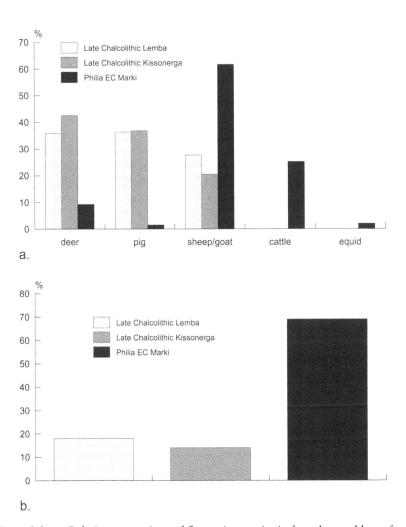

Figure 9.2 a. Relative proportions of five major species in faunal assemblages from Chalcolithic Lemba and Kissonerga and Philia EC Marki (after Webb & Frankel 2007: fig. 4); b. Relative proportions of rubbers and querns in the curated ground stone assemblages from Chalcolithic Lemba and Kissonerga and Philia EC Marki (drawn by the author from data in Elliott 1985, Elliott-Xenophontos 1998: table 19.2 and Frankel & Webb 2006: 219–30)

lower grinding stones) account for 69% of curated ground stone artefacts (Frankel & Webb 2006: 219), compared with 14% in Chalcolithic levels at Kissonerga (Elliott-Xenophontos 1998: table 19.2) and 18% at Lemba (Elliott 1985) (Figure 9.2b). This was accompanied by a twofold increase in the rubber-to-quern ratio and an increase in the size of grinding surfaces (Frankel & Webb 2006: 219). These changes in subsistence practices reflect different dietary preferences, human/animal relationships and work schedules and a significantly greater dependence on cereal processing in the EBA.

Other innovations that distinguish the EBA from the Chalcolithic cultural system include the systematic exploitation of local copper ores and an array of new mould-cast metal artefacts (including weapons); new agricultural technologies, such as backed sickle blades and the plough (replacing hoe-based agriculture); multi-roomed rectilinear stone and mud-brick architecture (replacing an 8000-year-old mono-cellular circular system); extramural burial and other aspects of mortuary behaviour and new ceramic wares, techniques and forms (for an extended discussion see Frankel et al. 1995; Webb & Frankel 1999, 2007). Other novelties fall within the domestic sphere. These include new hearth and oven types and related cooking equipment, new technologies of textile production (specifically low whorl spinning, the use of clay whorls and the vertical warp-weighted loom) and new childcare and household maintenance and discard practices (see, in particular, Webb & Frankel 2007). These imply major changes in modes of behaviour and related cultural practices, preferences, values and beliefs.

THE FINE-SCALE PERSPECTIVE: IDENTIFYING PROCESS

Modelling fine-scale process across this major disjunctive threshold is a far greater challenge. Expected correlates include a period of instability or destabilisation within Late Chalcolithic communities, as this well-established, highly adapted cultural system came under a new set of selective pressures when faced with the radically different operational environment of the Philia EC newcomers. Following Kirch's model (1980: 116, fig. 3.4), this is likely to have led to a period of low adaptedness (to the new environment) during which Late Chalcolithic communities may be expected to exhibit significant behavioural variation, as new and old practices coexisted and various change and response strategies were tried (Figure 9.3). This might be seen as a stress response or/and one involving a complex mix of competition, emulation and rejection. Chalcolithic communities were presented with a major adaptive challenge to which they had to respond radically rather than incrementally. New behaviour and new information needed to be recognised, learned and transmitted in some form *or* resisted and rejected as Philia settlements were progressively founded across the island. These adaptive pressures were not exerted by modifications in the real environment but by changes in the possibilities inherent in that environment. In this instance these involved a removal of factors that had previously set boundaries on

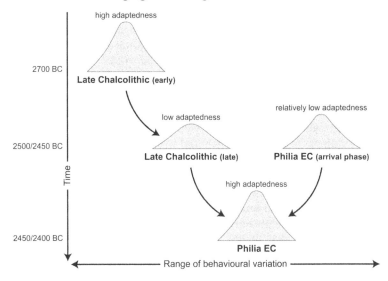

Figure 9.3   Predicted range of behavioural variability at times of high and low adaptedness in Late Chalcolithic and Philia EC communities over time (adapted from Kirch 1980: fig. 3.4)

the range of potential behaviour—for example, the constraints on population size and density and land use imposed by hoe-agriculture—and a major expansion in the social and economic choices available.

Some opportunities and choices would have been more attractive to Late Chalcolithic communities or easier to absorb, adopt or adapt than others. In some cases what was on offer were complex technologies (like metallurgy) or practices, like cattle breeding and training, which required specific expertise and acquired learning. Others simply involved different ways of doing things without immediately obvious advantages. The latter are likely to have included domestic practices. These typically entail habitual, small-scale, personal interactions between closely related individuals and involve deeply embedded cultural preferences and intergenerational expressions of family and community. In this domain, in particular, we might expect the interplay between social and technological choices to have been especially complex. Variation in the timing and nature of take-up processes may also be visible between internal and external domains of use (perhaps largely corresponding to female and male activities), between technologies and practices relating variously to the visible and nonvisible features of objects and to processes of transmission that involved implicit rather than explicit knowledge and experience (Pétrequin 1993; Lemonnier 1993).

Kissonerga is the only site with excavated exposures across the Late Chalcolithic/Philia EC transition. Located in the southwest, far distant from the earliest Philia EC settlements, it is likely to have coexisted with the incursive system for longer than most indigenous communities. Late Chalcolithic

Kissonerga is a two-phase period. Period 4a is dated to *c.* 2700–2500 BC; Period 4b is dated to the century from 2500 to 2400 BC, and thus to the generations immediately before the Late Chalcolithic was replaced at this site by the Philia EC in Period 5 (Peltenburg 1998: 249–58). It is therefore in Period 4b that we would expect to see the impact on this community of the coexisting Philia system. What we do see is of particular interest. In Period 4b there is no indication of Philia EC influence on domestic traditions, architecture or agricultural practices (see Webb & Frankel 2011). Mono-cellular circular buildings, traditional hearth and oven types and foodways continue unchanged. There are, however, a number of changes in burial practice (Figure 9.4). The Late Chalcolithic mortuary tradition typically involved single burial in intramural pits or shallow scoops (Peltenburg 1998: 68–73, fig. 4.1). Period 4b saw the limited introduction of the chamber tomb, a facility commonly found, alongside pit burials, in extramural cemeteries associated with the Philia EC. In another significant departure, the chamber tombs at Kissonerga were reserved primarily for adults and over half held multiple burials. Grave goods recovered from chamber tombs, however, belong exclusively to the Chalcolithic tradition (Peltenburg 1998: 71–2, fig. 4.1). Of interest also is a discrete group of chamber and pit tombs enclosed by a timber palisade (Peltenburg 1998: 46, 88–9, fig. 3.14). This unique mortuary facility is described by the excavator as a 'pseudo-cemetery' (Peltenburg 1998: 89).

Kissonerga Period 4b also produced evidence of small-scale copper processing, including several fragments of pure smelted copper, a piece of smelted ore and two stone dishes identified as possible crucibles (Peltenburg

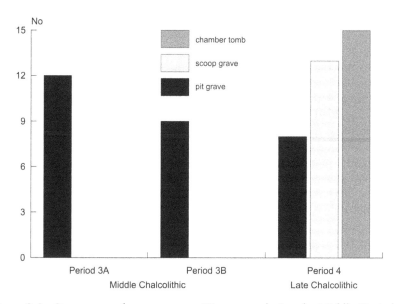

Figure 9.4   Occurrence of grave types at Kissonerga during the Middle (Period 3) and Late Chalcolithic (Period 4) (adapted from Peltenburg 1998: fig. 4.2)

1998: 188–9, fig. 95.14–15; 2011; on the context and date of this evidence see also Webb & Frankel 2011). Spurred annular pendants, a diagnostic Philia EC ornament, are also evident in Period 4b, and the presence of discards suggests on-site manufacture (Peltenburg 1998: 48–9, 190–91, figs. 97.29, 98.1, pl. 36.9). A small amount of other Philia material from Period 4b includes 51 sherds from Philia EC ceramic fabrics and a clay spindle whorl of a type closely paralleled on Philia EC settlements (Bolger 1998: 121; Peltenburg 1998: 199, fig. 100.19, pl. 37.9). Other diagnostic Philia elements are securely dated to Period 5. They include evidence of cattle, a fully Philia EC ceramic assemblage, jar burials for infants and a shift to extramural burial for adults. Although no structures survive, stone architecture is also indicated by remnant building material.

In sum, during Period 4b at Kissonerga domestic practices remained embedded in earlier traditions and there was no shift in architectural form, animal husbandry or agricultural technologies. Some people, however, were manufacturing Philia EC style pendants (alongside traditional ornament types), practising or attempting to practice small-scale metalworking and being buried in chamber tombs. Philia items, including pottery vessels, appear also to have occasionally reached Kissonerga from a Philia EC settlement or settlements nearby. This suggests that the inhabitants of Period 4b Kissonerga were interacting directly with Philia communities. An actual Philia presence, however, appears unlikely in view of the absence of any changes in domestic practice. Rather, the evidence suggests a differential take-up of elements of the Philia EC system by some Period 4b inhabitants. This adoption, manipulation or/and emulation of Philia material culture may have served as a mechanism of prestige enhancement in this Late Chalcolithic community; with some people deploying new fashions (pendants and chamber tombs) and taking up new knowledge (metallurgy) as part of an internal strategy of social exclusion. It may also have been a response to an increasingly destabilised Late Chalcolithic environment and to competition between neighbouring groups for regional authority or resources.

The evidence for metalworking and for the manufacture of Philia EC type pendants is associated in particular with a structure built in Period 4b directly over an impressive Period 4a building known as the Pithos House, which is thought to have been the residence of a group within Kissonerga who periodically attempted to exert authority (Peltenburg 1998: 254). While there is no direct association of this material with the Pithos House (see Webb & Frankel 2011 *contra* Peltenburg), the concentration of much of this evidence within a discrete area does suggest restricted access to these elements of Philia material culture and technology and possibly a link with earlier aggrandising elements in this community. The association of chamber tombs with particular grave goods, notably spouted flasks and faience beads, and the experimentation with discrete burial zones also provide some grounds for associating the adoption and recontextualisation of Philia practices with an elite segment of society (Peltenburg 1998: 91–2).

The emulation of 'foreign' practices in Kissonerga Period 4b certainly appears to have been selective. Some elements of the Philia EC system underwent a process of cultural negotiation and conversion, while others were either not known or actively rejected—specifically those related to the domestic domain where traditional habits, technologies and preferences continued unchanged from Period 4a. If dietary and cooking habits are sensitive indicators of shared community, the people of Period 4b Kissonerga and their Philia EC neighbours would appear to have remained distinct and there is little or nothing to suggest permanent movement between the two. There is no evidence here for a community-wide process of acculturation or hybridisation. Rather, technological and social choices appear to have been made in order to signal exclusiveness; and others rejected or ignored, perhaps to reinforce communal identity or simply because they were irreconcilable with the local reality and internal coherence of this Late Chalcolithic community. As predicted by Kirch's model, Kissonerga Period 4b shows greater diversity of material culture and practices than Period 4a, with the co-existence of varied arrangements, notably in burial habits (Figure 9.4), signalling a period of experimentation and stress (Kirch 1980: fig. 3.4). This diversity is described by the excavator as "completely novel for any restricted period of Cypriot prehistory" and likely to be linked to "a society in flux" (Peltenburg 1998: 89).

The processes involved in the transition to Philia EC Period 5 at Kissonerga are unfortunately obscured by poor preservation. Lack of evidence for widespread destruction of the latest Period 4 buildings or for a significant temporal gap led Peltenburg to suggest a peaceful transition without settlement interruption (1998: 52, 259). This may have involved people coming in, although there appears to have been a significant overall diminution in population. There is also some evidence for continuity in domestic practices, with *in situ* features suggesting that domestic installations typical of Period 4 were constructed alongside new types in Period 5 (Peltenburg 1998: 53, 258–9). The recycling of Period 4b vessels for Period 5 jar burials also suggests close connections with earlier traditions (Peltenburg 1998: 54; Bolger 2007: 179). It seems possible, therefore, that the Period 4b/5 transition involved the integration of Philia and indigenous communities or the encapsulation of one by the other. Beyond this chronological horizon the descendants of the Chalcolithic inhabitants lose archaeological visibility.

Kissonerga was a substantial settlement with a suggested population of 600 to 1500 people in Period 4b (Peltenburg 1998: 255). Its long history, size and distance from the earlier phases of Philia EC settlement may have allowed its inhabitants to maintain their cultural integrity in the co-presence of Philia communities with radically different technologies and lifeways for almost a century. Intercultural encounters seem specifically to have stimulated emulation, innovation and creativity, resulting in a number of hybrid practices (intramural chamber tombs and mortuary enclosures). Some discretionary elements reflect status enhancement, where new material was manipulated to serve older ends. There was less incentive to change other practices, especially in the domestic realm, where continuity and stability are clearly apparent. How processes of contact and integration played out elsewhere is unclear in the absence of excavated mid-third millennium indigenous settlements in other regions. It would appear, however, that the readjustments and choices made by Late Chalcolithic people across the island were cumulatively negative in the long term.

CONCLUSION

When Philia EC communities arrived on Cyprus in the mid-third millennium BC they brought a set of technologies and lifeways with them which presented a radical adaptive challenge to existing communities and put the latter at a major competitive disadvantage. These technologies appear to have been relatively rapidly 'mapped onto' a new physical environment, allowing the establishment of Philia EC settlements and the exploitation of local resources across much of the island. While this process no doubt involved multiple, complex and varied cultural encounters and a degree of hybridisation, it does not appear to have been an evenly bidirectional process and may have involved considerable cultural and social discrimination and confrontation. In attempting to understand the impact of the Philia EC migration, issues of timing, scale, process and cultural distance are clearly important. In seeking to understand the indigenous reaction, we can only propose a crude model which relates to one site (Kissonerga), and the response here may have been far from typical. Interaction between Philia EC and Chalcolithic people on the north coast, in particular, is likely to have involved different processes, and access to sites on the north coast (currently constrained by the political situation on the island) is a priority for future research.

At Kissonerga some elements of the Philia EC 'package' were adopted or emulated in order to signal exclusiveness; and others were rejected or not chosen, perhaps to reinforce cultural identity or simply because they were incomprehensible within the parameters of the perceived Late Chalcolithic environment. Over time competition for resources or primacy between communities is also likely to have added to the adoption imperative. The important point here is that within a constant or stable real environment, the choices made (and not made) appear to have been driven by socio-cultural factors rather than by technological imperatives. Technological choices were enabled or constrained by their perceived social 'value' and their relevance to existing activity and value systems rather than the relative inefficiency of existing practices. It may be that the Late Chalcolithic system 'failed' because the technological changes required for community-wide reciprocal and progressive adaptation or hybridisation involved a transformation in social relations and ideology that was irreconcilable with the internal coherence of Late Chalcolithic communities. The cognitive dimension is important; choices are made with regard to what is available, accessible and acceptable and may be suboptimal or even culturally maladaptive in the long term. It is this cognitive dimension (or the overlay of the perceived on the real environment) that is responsible for the fact that we cannot uniquely predict or model adaptive responses to exogenic change. In any case it is clear that the arrival of Philia EC communities led ultimately to the disintegration of the Chalcolithic cultural system in Cyprus and the marginalisation or encapsulation of most of its people.

REFERENCES

Ashmore, W. 2002. 'Decisions and dispositions': socializing spatial archaeology. *American Anthropologist* 104: 1172–83.

Bender, B. 2002. Time and landscape. *Current Anthropology* 43: S103–S112.

Bolger, D.L. 1991. Early Red Polished ware and the origin of the 'Philia Culture,' in J.A Barlow, D.L. Bolger & B. Kling (ed.) *Cypriot ceramics: reading the prehistoric record*: 29–36. Philadelphia: Leventis Foundation & University Museum of Archaeology and Anthropology, University of Pennsylvania.

Bolger, D.L. 1998. The pottery, in E. Peltenburg (ed.) *Lemba archaeological project II.1A: excavations at Kissonerga–Mosphilia, 1979–1992*: 93–147 (Studies in Mediterranean Archaeology 70:2). Jonsered: Paul Åströms förlag.

Bolger, D.L. 2007. Cultural interaction in 3rd millennium B.C. Cyprus: evidence of ceramics, in S. Antoniadou & A. Pace (ed.) *Mediterranean Crossroads*: 162–86. Athens: Pierides Foundation.

Broodbank, C. 2010. 'Ships a–sail from over the rim of the sea': voyaging, sailing, and the making of Mediterranean societies c. 3500–800 BC, in A. Anderson, J.H. Barrett & K.V. Boyle (ed.) *The global origins and development of seafaring*: 249–64. Cambridge: McDonald Institute for Archaeological Research.

Butzer, K.W. 1972. *Environment and archaeology: an ecological approach* (2nd edition). London: Methuen.

Croft, P. 1985. The mammalian fauna, in E. Peltenburg (ed.) *Lemba archaeological project I: excavations at Lemba Lakkous, 1976–1983*: 98–100, 202–208, 295–96 (Studies in Mediterranean Archaeology 70:1). Göteborg: Paul Åströms förlag.

Croft, P. 1998. Animal remains: synopsis, in E. Peltenburg (ed.) *Lemba archaeological project II.1A: excavations at Kissonerga–Mosphilia, 1979–1992*: 207–14 (Studies in Mediterranean Archaeology 70:2). Jonsered: Paul Åströms förlag.

Croft, P. 2006. The animal remains, in D. Frankel & J.M. Webb (ed.) *Marki Alonia. An Early and Middle Bronze Age settlement in Cyprus. Excavations 1995–2000*: 263–81 (Studies in Mediterranean Archaeology 123.2). Jonsered: Paul Åströms förlag.

Dikaios, P. 1962. The Stone Age, in P. Dikaios & J.R. Stewart (ed.) *Swedish Cyprus Expedition* Volume IV. Part IA: 1–204. Lund: The Swedish Cyprus Expedition.

Dikomitou, M. 2010. A closer look at Red Polished Philia fabrics. Inquiring into ceramic uniformity in Cyprus, ca. 2500–2300 B.C. *The Old Potter's Almanack* 15: 1–6.

Dikomitou, M. 2011. Ceramic production, distribution, and social interaction. An analytical approach to the study of Early and Middle Bronze Age pottery from Cyprus. Unpublished PhD dissertation, University College London.

Dikomitou, M. & M. Martinon-Torres. 2012. Fabricating an island-wide tradition. Red Polished pottery from Early and Middle Bronze Age Cyprus, in N. Zacharias, M. Georgakopoulou, K. Polikreti, Y. Facorellis & T. Vakoulis (ed.)

Proceedings of the 5th HSA symposium (October 2008 Athens): 423–42. Athens: Papazisi Publishers.

Dobres, M.-A. 2000. *Technology and social agency: outlining an anthropological framework for archaeology.* Oxford: Blackwell.

Dobres, M.-A. 2001. Meaning in the making: agency and the social embodiment of technology and art, in M.B. Schiffer (ed.) *Anthropological perspectives on technology:* 47–76. Albuquerque: University of New Mexico Press.

Efe, T. 2007. The theories of the 'Great Caravan Route' between Cilicia and Troy: the Early Bronze Age III period in inland Western Anatolia. *Anatolian Studies* 57: 47–64.

Elliott, C. 1985. The ground stone industry, in E. Peltenburg (ed.) *Lemba archaeological project I: excavations at Lemba Lakkous, 1976–1983*: 70–93, 161–95, 271–75 (Studies in Mediterranean Archaeology 70:1). Göteborg: Paul Åströms förlag.

Elliott–Xenophontos, C. 1998. Ground stone tools, in E. Peltenburg (ed.) *Lemba archaeological project II.1B: excavations at Kissonerga–Mosphilia, 1979–1992*: 193–232 (Occasional Paper 19). Edinburgh: Department of Archaeology, University of Edinburgh.

Frankel, D. 2000. Migration and ethnicity in prehistoric Cyprus: technology as habitus. *European Journal of Archaeology* 3: 167–87.

Frankel, D. & J.M. Webb. 2004. An Early Bronze Age shell pendant from Cyprus. *Bulletin of the American Schools of Oriental Research* 336: 1–9.

Frankel, D. & J.M. Webb. 2006. *Marki Alonia. An Early and Middle Bronze Age settlement in Cyprus. Excavations 1995–2000* (Studies in Mediterranean Archaeology 123:2). Jonsered: Paul Åströms förlag.

Frankel, D., J.M. Webb & C. Eslick. 1995. Anatolia and Cyprus in the third millennium BCE. A speculative model of interaction. *Abr-Nahrain Supplement* 5: 37–50.

Georgiou, G., J.M. Webb & D. Frankel. 2011. *Psematismenos-Trelloukkas. An Early Bronze Age cemetery in Cyprus.* Nicosia: Department of Antiquities, Cyprus.

Hennessy, J.B., K.O. Eriksson & I.C. Kehrberg. 1988. *Ayia Paraskevi and Vasilia. Excavations by J.R.B. Stewart* (Studies in Mediterranean Archaeology 82). Göteborg: Paul Åströms förlag.

Kirch, P. 1980. The archaeological study of adaptation: theoretical and methodological issues, in M.B. Schiffer (ed.) *Advances in Archaeological Method and Theory* 3: 101–56. New York: Academic Press.

Knapp, A.B. 2008. *Prehistoric and protohistoric Cyprus. Identity, insularity and connectivity.* Oxford: Oxford University Press.

Knapp, A.B. & W. Ashmore. 1999. Archaeological landscapes: constructed, conceptualized, ideational, in W. Ashmore & A.B. Knapp (ed.) *Archaeologies of landscape: contemporary perspectives*: 1–31. Oxford: Blackwell.

Lemonnier, P. 1993. Introduction, in P. Lemonnier (ed.) *Technological choices: transformation in material cultures since the Neolithic*: 1–35. London & New York: Routledge.

Manning, P. 2005. *Migration in world history.* London & New York: Routledge.

Manning, S.W. 1993. Prestige, distinction, and competition: the anatomy of socioeconomic complexity in fourth and second millennium B.C.E. Cyprus. *Bulletin of the American Schools of Oriental Research* 292: 35–58.

Manning, S.W. & S. Swiny. 1994. Sotira Kaminoudhia and the chronology of the Early Bronze Age in Cyprus. *Oxford Journal of Archaeology* 13: 149–72.

Peltenburg, E. (ed.) 1985. *Lemba archaeological project I: excavations at Lemba–Lakkous, 1976–1983* (Studies in Mediterranean Archaeology 70:1). Göteborg: Paul Åströms förlag.

Peltenburg, E. 1998: *Lemba archaeological project II.1A: excavations at Kissonerga–Mosphilia, 1979–1992* (Studies in Mediterranean Archaeology 70:2). Göteborg: Paul Åströms förlag.

Peltenburg, E. 2007. East Mediterranean interaction in the 3rd millennium BC, in S. Antoniadou & A. Pace (ed.) *Mediterranean crossroads*: 139–59. Athens: Pierides Foundation.

Peltenburg, E. 2011. Cypriot Chalcolithic metalwork, in P.B. Betancourt & S.C. Ferrence (ed.) *Metallurgy: understanding how, learning why: studies in honor of James D. Muhly*: 3–10. Philadelphia: INSTAP Academic Press.

Pétrequin, P. 1993. North wind, south wind. Neolithic technical choices in the Jura Mountains, 3700–2400 BC, in P. Lemonnier (ed.) *Technological choices: transformation in material cultures since the Neolithic*: 36–76. London & New York: Routledge.

Şahoğlu V. 2005. The Anatolian Trade Network and the Izmir region during the Early Bronze Age. *Oxford Journal of Archaeology* 24: 339–61.

van der Leeuw, S.E. 1993. Giving the potter a choice: conceptual aspects of pottery techniques, in P. Lemonnier (ed.) *Technological choices: transformations in material cultures since the Neolithic*: 238–88. London & New York: Routledge.

Webb, J.M. 2002. New evidence for the origins of textile production in Bronze Age Cyprus. *Antiquity* 76: 364–71.

Webb, J.M. & D. Frankel. 1999. Characterizing the Philia facies. Material culture, chronology and the origin of the Bronze Age in Cyprus. *American Journal of Archaeology* 103: 3–43.

Webb, J.M. & D. Frankel. 2007. Identifying population movements by everyday practice. The case of third millennium Cyprus, in S. Antoniades & A. Pace (ed.) *Mediterranean crossroads*: 189–216. Athens: Pierides Foundation.

Webb, J.M. & D. Frankel. 2008. Fine ware ceramics, consumption and commensality: mechanisms of horizontal and vertical integration in Early Bronze Age Cyprus, in L.A. Hitchcock, R. Laffineur & J. Crowley (ed.) *Dais. The Aegean feast. Proceedings of the 12th International Aegean Conference. University of Melbourne, Centre for Classics and Archaeology, 25–29 March 2008*: 287–95. (Aegaeum 29). Liège: Université de Liège & University of Texas at Austin.

Webb, J.M. & D. Frankel. 2011. Hearth and home as community identifiers in Early Bronze Age Cyprus, in V. Karageorghis & O. Kouka (ed.) *About cooking pots, drinking cups, loom weights and ethnicity in Bronze Age Cyprus and the neighbouring regions*: 29–42. Nicosia: Leventis Foundation.

Webb, J.M., D. Frankel, Z.A. Stos & N. Gale. 2006. Early Bronze Age metal trade in the Eastern Mediterranean. New compositional and lead isotope evidence from Cyprus. *Oxford Journal of Archaeology* 25: 261–88.

10 Landscape Learning in Colonial Australia
Technologies of Water Management on the Central Highlands Goldfields of Victoria

Susan Lawrence and Peter Davies

INTRODUCTION

This chapter explores intersections between a number of related themes: between environment and technology, but also between environmental and landscape archaeology and between archaeological and documentary evidence. Our aim is to draw on all of these themes to gain a better understanding of human–environmental interactions in the recent past, using technological responses to water scarcity during Australia's nineteenth-century gold rush as a case study. Historical archaeology provides the disciplinary context for integrating archaeological and documentary evidence, while a landscape learning model provides the framework for integrating environmental and landscape archaeology. Our study area is the Central Highlands district of the Australian state of Victoria, a region that was the scene of one of the world's great nineteenth-century gold rushes. The region has a highly variable annual rainfall, generally concentrated in the winter and spring, while alternating cycles of drought and flood presented challenges for the water-intensive mining industry. The study of mining has been a mainstay of historical archaeology in Australia since it began in the 1970s, although specific study of the environment has been more muted. Here we invert the normal relationship, making the environment the focus of research, rather than the background.

Landscape learning is the process by which individuals and groups learn about new environments and incorporate this new information into ongoing cultural systems (Rockman 2003, 2010). It is particularly relevant in contexts of colonisation and migration but also where there are significant changes to environmental regimes (e.g. climate change) or resource exploitation strategies (e.g. from pastoralism to mining). Landscape learning incorporates three types of information gathered over varying periods of time as a group becomes familiar with its new environment and as adaptation takes place. The first type of information, locational knowledge, concerns the characteristics of the new environment, such as the spatial distribution and quality of different plant, animal and mineral resources, and information about seasonal temperature and rainfall variation. This information can be gathered quickly within days or weeks, or over a longer time as new require-ments are identified. The second type of information is limitational knowl-edge, concerning the quality and reliability of the resource. Depending on the nature of the data, this can take a much longer period to acquire as the resources are put to use. Information about the full range of variability with regard to climate and animal populations, in particular, can take years, decades and even generations to accumulate as full cycles are experienced. The final type of information is social knowledge, the collective experience, knowledge and tradition that enables the group to interact successfully with the environment and that creates a sense of place.

As a model for understanding how adaptation to new environments takes place, landscape learning has the advantage of integrating science-based environmental data with the more social and cognitive approaches of landscape archaeology. Although landscape studies have long had a place in historical archaeology, there has been a failure to engage fully with en-vironmental data or to recognise the potential of these studies to contrib-ute to understandings of the dominant themes in the archaeology of the modern world, including colonisation, urbanisation and industrialisation (Kealhofer 1999; Hardesty 1999, 2009; Mrozowski 2006, 2010; Deagan 2008; Branton 2009; Bain 2010). In part this has been due to the different origins and trajectories of environmental and historical archaeology, as the former, with its links to the biological and earth sciences, has embraced more positivist epistemologies while the latter's engagement with history and the humanities has been more influenced by post-modernist scholar-ship (Deagan 2008). Mrozowski (2006) also points to the abiding interest in material culture that has strongly characterised historical archaeology, cutting across the otherwise diverse and sometimes competing theoretical perspectives of practitioners that have encompassed structuralism, Marxism and post-processualism.

The situation described above is also broadly true in Australian histori-cal archaeology. While there have been a number of important environ-mental archaeological studies in relation to colonisation (Gale & Haworth 2002; Gale et al. 2004; Tibby 2004), industrialisation (Lentfer et al. 1997) and urbanisation (Macphail 1999; Macphail & Casey 2009), landscape ap-proaches have been much more dominant. Australian historical archaeology has benefited from a long engagement with cultural geography (e.g. Meinig 1962; Powell 1970; Jeans 1972; Williams 1974). One of the earliest inter-pretive frameworks proposed in Australian historical archaeology was the 'Swiss Family Robinson' model, developed by archaeologist Judy Birming-ham together with geographer Dennis Jeans, a model in which environmental learning and adaptation at the landscape level was central (Birmingham & Jeans 1983). More recently historical archaeologists have also begun to engage with the innovative work of Australian environmental historians who demonstrate how history and the environment can be brought together at the global scale (e.g. Griffiths & Robin 1997; Tyrrell 1999; Griffiths 2001; McKernan 2005; Sherratt et al. 2005; Robin 2007; Clode 2010), but the influence of this scholarship in the humanities has tended to draw attention away from the natural sciences.

Landscape approaches, with their links to social theory, have the appearance of being a more natural fit with the interests of historical archaeology, and in that context it is not surprising that historical archaeologists have not responded more enthusiastically to Kealhofer's (1999) appeal for the closer integration of environmental and landscape approaches and the use of environmental data as evidence for meaning and behaviour. Kealhofer (1999: 370) maintains that landscapes are not simple artefacts of human activity (as per Deetz 1990) but are dynamic, changing palimpsests of information at different scales and of different types. Because of this, Kealhofer argues, "the impact and meaning of places and spaces *is* strongly interwoven with the way places are both materially and biologically composed" (Kealhofer 1999: 381, emphasis in original). Landscape learning is one way of attempting to bridge this divide, as environmental archaeology, the "biological dimension of cultural processes" (Hardesty 2009: 67), is necessary to provide information about locational and limitational factors, while landscape approaches, with their concern for understanding how people "conceptualised, organised, and manipulated their environments" (Branton 2009: 51), provide the structure for incorporating social factors. As Rockman (2003: 7) states, "[social information] linking environmental information and social practice can be integral to a group's sense of place and the creation of a cultural landscape."

WATER SCARCITY AND THE AUSTRALIAN GOLD RUSH

In February 1851 prospectors fresh from the California gold rush discovered gold in New South Wales. Even richer discoveries soon followed in the adjacent colony of Victoria, triggering a massive migration of people that transformed the society, economy and landscape of what had been a small and insignificant colonial outpost of the British Empire. Within 10 years of the discovery of gold the Victorian population had grown from 77,000 to 540,000 and more than 25 million ounces of gold, about one-third of total world output, had been recovered from the soil (Serle 1963: 382–90). While most of the migrants were from the British Isles and elsewhere in Australia, a considerable number were from other European countries, including France, Germany, Switzerland and Italy, the United States and Canada, and the Pacific Islands. Significantly, the largest non-British contingent was around 40,000 migrants (almost exclusively male) from southern China.

Most of the gold discoveries were in the mountainous region running across central Victoria from the Western District eastwards into Gippsland

and the southern parts of the Great Dividing Range, which runs along the east coast of Australia (Figure 10.1). These inland areas had been thinly settled by white pastoralists for 15–20 years and by Aboriginal people for some 40,000 years before that. As a result of the gold discoveries new settlements appeared almost overnight. Some disappeared just as quickly but others, such as Ballarat and Bendigo, grew to substantial cities of tens of thousands of people. The diggings stimulated the need for local manufacturing of consumer goods and mining machinery; infrastructure such as roads, railways and telegraph lines; agriculture and horticulture to provide food; and timber getting for the mines and towns. Non-mining land was rapidly taken up for farming as sheep graziers lost their dominance, while the region's Aboriginal people, already displaced by pastoralism, were pushed even further to the fringes.

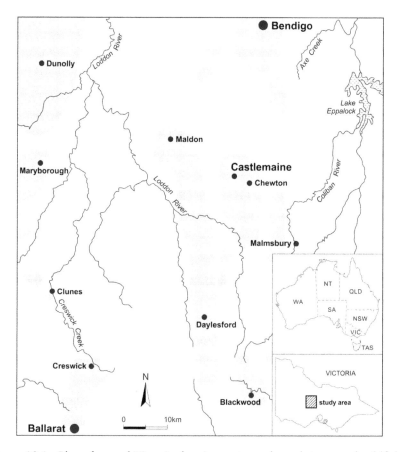

Figure 10.1   Plan of central Victoria showing main creeks and rivers, and goldfields in grey

Mining placed heavy demands on many natural resources, and while some, such as timber, were readily available, it quickly became apparent that water was more of a problem. Many of the diggings in central Victoria were located on the northern slopes of the Great Dividing Range, with a drier rainfall pattern than districts to the south (Lee 1982: 27–30). As the miners discovered, the region is characterised by a Mediterranean climate with highly variable rainfall coming mainly in the winter and spring months (June to October), and generally hot and dry summers. The region is also subject to significant fluctuations in annual rainfall (Figure 10.2), along with a high rate of surface evaporation and irregular groundwater supplies. All these factors placed a premium on rainfall and natural flows in creeks and rivers.

Rainfall and surface water in Victoria may be seasonally and cyclically in short supply, but mining demanded large quantities of water year-round. Despite its scarcity, miners were profligate in their use of water, using large quantities to wash gold from earth and discharging the sludge into streams, causing rapid siltation that contributed to damaging floods. In Bendigo in 1858, for example, 10,000 men and 5000 horses worked 2000 puddling machines (all of which relied on water to operate) in order to separate gold from dirt (Powell 1989: 49). Later water-intensive mining techniques, including hydraulic sluicing and dredging, wrought environmental havoc to entire river valleys, the effects of which are still easily seen today (McGowan 2001). Miners constructed extensive networks of water races to channel water from one location to another. The sheer scale of these races was recorded by Robert Brough Smyth, who served as Secretary for Mines in Victoria from 1860 to 1876 (1980: 547). He calculated that more than 2400

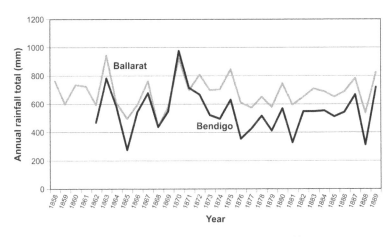

Figure 10.2 Annual rainfall recorded at Ballarat Survey Office (station #089048) and Bendigo Prison (station #081003) from 1858 to 1889.

Source: http://www.bom.gov.au/climate/data (accessed 11 February 2011)

miles (3900km) of races had already been constructed in Victoria's principal mining districts by 1868, a figure that does not include the small-scale water systems created for watering farms, market gardens and local townships, and that reflects a period before the large-scale advent of hydraulic sluicing.

As settlements grew, water was also required for market gardening and for domestic consumption, and then for agriculture. Chinese market gardeners used water in a way that was apparently stable and productive, and that hinted at the fertility and productivity that could be achieved through intensive cultivation, communal labour and careful water management (Frost 2002). However, the large urban populations on the goldfields increased pressure on the available water. The failure of domestic supplies to Sandhurst (Bendigo) during the drought of 1865 precipitated a crisis that eventually resulted in the government construction of the Coliban System of Waterworks (Russell 2009), while in Ballarat local government and miners funded and then wrangled over schemes to divert water from nearby watersheds (Nathan 2007). The 1880s saw major developments in irrigated agriculture in Victoria, with weirs built at public expense on the Loddon and Goulburn rivers, establishing the core of twentieth-century irrigation facilities. By 1896 there were several hundred thousand acres of farmland under irrigation in the northern plains, producing wheat, lucerne and sown grasses (Powell 1989: 117).

Miners, farmers and business people responded to water scarcity in a variety of ways. Some looked to the state and agitated for greater intervention, which resulted in infrastructure like the Coliban System and the enactment of legislation to regulate access to water, including the provisions of the Mining Statute 1865 (Brough Smyth 1980; Powell 1989; Russell 2009). More immediately, individuals, small groups and entrepreneurs built thousands of dams, culverts and races to store and move water. In doing so they created a cultural landscape of water management, transforming the environment to provide for human needs. This landscape is manifested in large- and small-scale local constructions and land transformations that are integrated over a wide area in linear features that connect distant dams and water sources to the mines, homes and gardens they were built to serve. It is also revealed in biological and geological data that reveal erosion, deforestation and sedimentation, the latter two effects sometimes at considerable distances from mining areas.

STUDY AREA

Mining occupied a large proportion of Victoria's land surface in the nineteenth century and water management issues were universal, although the details varied according to local topography, rainfall and mineralogy. Our analysis here focuses on part of the catchment of the upper Loddon River and the associated gold reefs of the region (Figure 10.1). The Loddon drains part of the northern slopes of the Great Dividing Range and flows into the Murray River. The region includes rugged mountainous uplands, river valleys and grassy plains, including extensive box–ironbark forests and stands of river red gums. Today the population is concentrated around the major centres of Bendigo and Castlemaine. In the second half of the nineteenth century these towns were the centres of two of Australia's major goldfields, the Mount Alexander (Castlemaine) and Sandhurst (Bendigo) fields, between them responsible for producing more than 28 million ounces (794 tonnes) of gold (Castlemaine Pioneers 1972: ix; Russell 2005: 115). Both had enormous deposits of easily recovered gold in surface alluvial gravels. Bendigo was also found to have even richer deposits in deeply buried quartz reefs, resulting in what were then some of the world's deepest underground mines and providing the basis for a large urbanised population. Around Castlemaine the quartz reefs, while rich, were shallower and more quickly worked out, and subsequent development has been minimal, resulting in the survival into the present of a unique cultural landscape partially protected in the Castlemaine Diggings National Heritage Park and currently being considered for nomination to the World Heritage List.

Our preliminary research has identified more than 600 previously recorded archaeological sites that contain features related to water diversion and storage in the Central Highlands area. These are primarily associated with mining and include dams of various sizes, water races, puddling sites, the remains of flumes, diversion tunnels, waterwheel foundations and sludge ponds. Domestic sites include wells, cisterns, dams and water tanks, while local and municipal water works, some of which remain in operation today, include diversion dams, stone-lined channels and aqueducts, spillways and standpipes. Mining activity, settlement and agriculture in the region have thus created a complex landscape of water management. The diverse and frequently well-preserved archaeological features, distributed within a highly modified natural environment, embody vital information about the landscape learning process that created them.

LEARNING ABOUT GOLD AND WATER

While most of the miners were new migrants entirely unfamiliar with the Australian environment, even those with colonial experience were using the land and its resources in a totally different way from the earlier pastoralists. The landscape learning process was, therefore, relevant to all. Locational knowledge, and particularly the location and nature of gold deposits, was vital. In some early instances gold was visible on the ground surface, particularly after rain, or glittering in quartz outcrops, but for the most part miners learned about the geology by trial and error, following the alluvial gold deeper underground and digging surface trenches to intercept buried quartz reefs. Based in large part on information derived from the mines, it is now known that the primary gold originates in elongated quartz reefs or veins running in a north–south direction, emplaced within folded and faulted sedimentary layers that once formed part of the continental shelf of Gondwana (Weston 1992). Costeans, or prospecting trenches, are still visible today running in an east–west direction, indicating that the miners soon learned to recognise the lines of the outcrops and how to most effectively locate them.

Where to find water was equally crucial. Indispensable for life itself, water was also needed to work the machines and wash gold from the soil and rock. The location of mining settlements, however, was based above all else on the location of gold, with the availability of other environmental resources a distant second. In Victoria the geological events that emplaced the gold occurred in the Ordovician period around 440 million years ago. Subsequent erosion and volcanic activity, the latter over the past 5 million years, have changed the landscape considerably since then, and modern topography and surface hydrology is only distantly related to the location of gold (Wilkinson 1988; Phillips & Hughes 1998). A map of the goldfields towns and the rivers in the study area (Figure 10.1) highlights the problem—the major gold deposits are generally located between the main branches of the rivers, not along them. This is in direct contrast to the California gold districts, where the gold had been eroded out of the host rock more recently and was found in and along the Quaternary river systems, providing abundant and accessible water for consumption and mining use. Compounding the problem in Victoria is the absence of surface water in lakes and marshes, and the comparatively low volume of flow in rivers.

Archaeological sites and features show the steps taken by miners to harvest rainwater. On domestic sites run-off from roofs was diverted into above-ground tanks, including disused wooden casks and old cast-iron ships' tanks, and later into underground cisterns, beehive shaped and lined with bricks and plaster. Mining sites show similar strategies on larger scales. Dams and reservoirs are common features where surface run-off has been stored and where water from puddlers has been recycled in settling ponds. Some facilities were limited in scale and built on-site as part of the mining operation to provide water for puddling machines, steam engines and stamp batteries. Others were large-scale commercial dams with lengthy race networks that provided water to paying customers tens of kilometres away. By the late 1860s more than 35 reservoirs had been constructed in the goldfields district, some containing more than 300 megalitres of water (Brough Smyth 1980: 401).

Limitational knowledge about mineralogy was obtained more slowly, as miners struggled to separate the gold from quartz and other minerals and sulphides. The goldmining landscape around Castlemaine is particularly informative about the process of acquiring limitational knowledge. Trial and error is evident in the abandoned workings and the remains of ineffective technologies, such as the foundations of stamp batteries that failed to extract payable gold from sulphide-rich ore, and waterwheels that failed to generate enough power. Success and improved technology are evident in the reworking of deposits and tailings dumps, where hydraulic sluicing and then dredging has destroyed all but the fringes of earlier ground sluicing sites, shallow sinkings and puddlers, and where tailings sands from stamp batteries have been removed and treated again in cyanide tanks. Mining sites thus represent complex landscapes that embody the process of learning about the environment.

Knowledge about climatic limitations could only be obtained through observation over seasons, years and decades. Miners quickly identified, however, the seasonal cycle of rain and dry periods and the characteristics of different fields. Within the first 12 months Bendigo was plagued by drought and floods, and it was soon known to be a 'winter' diggings where washdirt was stored in heaps on claims until the rains came and it could be worked (Russell 2005: 115). People were much slower to recognise the longer-term oscillations in drought and flood, and it was several decades before these were accepted as regular occurrences. In the decades before the completion of the Coliban System of Waterworks in 1877, prolonged summer droughts occurred in 1865 and 1868, and high-rainfall events that produced winter flooding took place in 1863, 1870 and 1875 (Figure 10.2). Arrangements to deal with drought and flood took time to complete. The Coliban eventually provided a stable supply of water but working out how much water it was necessary to store and where it would come from was a significant problem for those constructing the system. Accurate, long-term rainfall data were needed, along with reliable mapping of the catchments, but even by 1871 these were not available, as regular recording of rainfall data in the region did not commence until 1862, more than 20 years after initial white settlement (Russell 2009: 178).

Locational and limitational knowledge represent learning about the physical environment. Social knowledge incorporates the cultural capital and experience brought to that environment, and the processes of adapting to what is found. Technology may be part of cultural capital and may also be developed in the course of adaptation. The miners were hampered from the outset by the contrast between the climate of their homelands and that of the diggings. British colonists were "wet-country people" (Cathcart 2009: 259) whose ideals and expectations of water, fertility and the natural world generally could not easily be accommodated in the drier environmental reality of Australia. Like the Europeans, Chinese miners came to Australia from a much wetter environment, the lush and fertile south-eastern provinces of Guangdong and Fujian, and brought with them the idea that water would be readily available. Some of the technical knowledge the migrants brought with them was, therefore, inappropriate. Waterwheels, traditionally used for power in Britain, were tried at a number of mines in Victoria but success was often limited because of the variability of water flows (Lawrence et al. 2000). Chinese miners in California used their knowledge of irrigation technology to introduce a pumping system that relied on bucket bailers and waterwheels. On the American diggings the system was so successful it was called the "Chinese pump" and its use became standard (Hardesty 2003: 90), but it was used much less frequently in Victorian mining.

Other technologies for working with water proved more adaptable. Miners from the copper and tin areas of Cornwall and Devon contributed a great deal of expertise to the Australian mines, including the use of puddlers. A puddler was a circular trough dug in the ground, into which washdirt and water were placed. A horse tethered to a pole in the centre of the trough then walked around the exterior, drawing rakes through the water to loosen the clay and gravel and separate the gold. Puddlers were ideally suited to the heavy clay soils and to the challenges of water scarcity. They used relatively little water, which was harvested on-site in small dams and then captured for later reuse in settling tanks. Thousands of puddlers were used on the Bendigo and Mt Alexander goldfields, but they are little known outside Australia and New Zealand (Davey 1996: 53–4; Ritchie & Hooker 1997: 7). While California miners puddled clays in wooden tubs, the technology was probably derived from alluvial tin mining in Devon, where silts and clays were "buddled" in large water-filled boxes (Thorndycraft et al. 2003: 19).

Another technology that was embraced by all was the use of reservoirs (dams in the Australian vernacular) to store and equalise water supply in an attempt to transcend locational and limitational factors in the environment (Figure 10.3). The drought of 1865–66, in particular, stimulated a number

Figure 10.3   Small holding dam at Humbug Hill, near Creswick in central Victoria (Photo: S. Lawrence 2011)

Figure 10.4 Water race to Humbug Hill, near Creswick in central Victoria (Photo: P. Davies 2011)

of companies in the Bendigo region to construct their own dams, so that by 1871 at least 206 private dams were in use (Russell 2009: 174). Where sufficient storages could not be built on-site, races and dams had to be built to divert water from where it was available to where it was needed. In 1859, for example, the Humbug Hill Sluicing Company built a dam of up to 20 million gallons (90ML) capacity on Creswick Creek, connected by a 10km water race to the site of the group's mining claim (Figure 10.4). In the following year the company began extending the race another 10km in order to sell water to miners whose supplies were failing. The race was conveyed 800m across a nearby creek through a patent bitumen pipe (*Creswick Advertiser* 1862). By this stage the company controlled one of the most significant water privileges in the district, and when alluvial mining in the area declined in the 1870s the dam, races, piping and other features were sold to the local council to ensure the water supply of town residents.

CONCLUSION

Technology is part of the social knowledge used to develop solutions to locational and limitational challenges in the environment. In the case of the central Victorian goldfields, technology was used to optimise what was available in the environment, making use of scarce water to exploit abundant gold. The provision of a secure water supply enabled mining to progress and transcended the variability of climate. One of the unforeseen consequences, however, was widespread environmental damage (McGowan 2001). The manipulation of water in the service of mining caused localised erosion at the mines and in the adjacent streams and gullies, while heavy sediment loads choked the downstream waterways, altered their courses and caused periodic flooding (Parliament of Victoria 1887). The Humbug Hill Sluicing Company was one of dozens of alluvial mining parties in the area that released great quantities of sludge into Creswick Creek. Dredging operations around 1900 and further sluicing in the 1930s exacerbated the problem. The dynamics of the creek's original hydrology have been so severely disrupted by mining that even today the town of Creswick is one of the worst flood-affected places in Victoria during extreme rainfall periods such as those of 2010–11, when it was flooded four times.

Although the consequences of technology use have not been a focus here, it is clear that the use of technology can and does significantly alter the environment, creating a feedback loop with locational and limitational knowledge as conditions change. While 'knowledge' implies a static situation, landscape learning is in practice an ongoing process. In the central Victorian goldfields, the application of traditional technologies to overcome the limits of water scarcity created new possibilities for mining, in effect a new environment, and this in turn created new problems of erosion and sedimentation. The Coliban System was intended to provide water for mining purposes at Castlemaine and Bendigo, but by the time it was completed companies with puddling machines and stamp batteries had largely made their own arrangements regarding water supplies. Instead, the Coliban was used for a new kind of mining—hydraulic sluicing, or hydraulicking—a system that involved channelling water downhill through iron pipes and then through a nozzle that directed the high pressure stream at a cliff-face, undermining the deposit and washing out the gold. With abundant water now available, the region became a centre for hydraulic sluicing in Victoria (Russell 2009: 212). The devastation to the environment increased, with the practice denuding hillsides of soil down to the bedrock and blasting new cliffs and valleys, while the sediment that was washed out filled up the gullies and created new deposits that were reworked by dredges in the 1930s. In central Victoria mining has been a significant intervention in the environment, creating and re-creating the landscapes of geology and water.

REFERENCES

Bain, A. 2010. Environmental and economic archaeologies of missions, colonies, and plantations. *Historical Archaeology* 44(3): 1–3.
Birmingham, J. & D. Jeans. 1983. The Swiss Family Robinson and the archaeology of colonisations. *Australian Journal of Historical Archaeology* 1: 3–14.

Branton, N. 2009. Landscape approaches in historical archaeology: the archaeology of places, in T. Majewski & D. Gaimster (ed.) *International handbook of historical archaeology*: 51–66. New York: Springer.

Brough Smyth, R. 1980. *The goldfields and mineral districts of Victoria*. Melbourne: Queensberry Hill Press.

Castlemaine Pioneers. 1972. *Records of the Castlemaine Pioneers*. Adelaide: Rigby.

Cathcart, M. 2009. *The water dreamers: the remarkable history of our dry continent*. Melbourne: Text Publishing.

Clode, D. 2010. *Future in flames*. Melbourne: Melbourne University Press.

Creswick Advertiser. Mining news, 3 June 1862: 2.

Davey, C. 1996. The origins of Victorian mining technology, 1851–1900. *The Artefact* 19: 52–62.

Deagan, K. 2008. Environmental and historical archaeology, in E. Reitz, C.M. Scarry & S. Scudder (ed.) *Case studies in environmental archaeology*: 21–42. New York: Springer.

Deetz, J. 1990. Prologue: landscapes as cultural statements, in W. Kelso & R. Most (ed.) *Earth patterns: essays in landscape archaeology*: 1–4. Charlottesville: University Press of Virginia.

Frost, W. 2002. Migrants and technological transfer: Chinese farming in Australia. *Australian Economic History Review* 42(2): 113–31.

Gale, S.J. & R.J Haworth. 2002. Beyond the limits of location: human environment disturbance prior to official European contact in early colonial Australia. *Archaeology in Oceania* 37: 123–36.

Gale, S.J., R.J. Haworth, D.E. Cook & N.J. Williams. 2004. Human impact on the natural environment in early colonial Australia. *Archaeology in Oceania* 39(3): 148–56.

Griffiths, T. 2001. *Forests of ash: an environmental history*. Cambridge: Cambridge University Press.

Griffiths, T. & L. Robin. 1997. *Ecology and empire: environmental history of settler societies*. Melbourne: Melbourne University Press.

Hardesty, D. 1999. Historical archaeology in the new millennium: a forum. *Historical Archaeology* 33(3): 51–8.

Hardesty, D. 2003. Mining rushes and landscape learning in the modern world, in M. Rockman & J. Steele (ed.) *Colonization of unfamiliar landscapes: the archaeology of adaptation*: 81–95. London: Routledge.

Hardesty, D. 2009. Historical archaeology and the environment: a North American perspective, in T. Majewski & D. Gaimster (ed.) *International handbook of historical archaeology*: 67–76. New York: Springer.

Jeans, D.N. 1972. *An historical geography of New South Wales to 1901*. Sydney: Reed.

Kealhofer, L. 1999. Adding content to structure: integrating environment and landscape, in G. Egan & R. Michael (ed.) *Old and New Worlds*: 378–89. Oxford: Oxbow.

Lawrence, S., K. Hoey & C. Tucker. 2000. Archaeological evidence of ore processing on the Howqua goldfield. *The Artefact* 23: 15–21.

Lee, D.M. 1982. Climate, in J.S. Duncan (ed.) *Atlas of Victoria*: 26–31. Melbourne: Victorian Government Printing Office.

Lentfer, C., W. Boyd & D. Gojak. 1997. Hope Farm windmill: phytolith analysis of cereals in early colonial Australia. *Journal of Archaeological Science* 24: 841–56.

Macphail, M. 1999. A hidden cultural landscape: colonial Sydney's plant microfossil record. *Australasian Historical Archaeology* 17: 79–109.

Macphail, M. & M. Casey. 2009. 'News from the Interior': What can we tell from plant microfossils preserved on historical archaeological sites in colonial Parramatta? *Australasian Historical Archaeology* 26: 45–70.

McGowan, B. 2001. Mullock heaps and tailing mounds: environmental effects of alluvial goldmining, in I. McCalman, A. Cook & A. Reeves (ed.) *Gold: forgotten histories and lost objects of Australia*: 85–102. Cambridge and Melbourne: Cambridge University Press.

McKernan, M. 2005. *Drought: the red marauder*. Sydney: Allen & Unwin.

Meinig, D.W. 1962. *On the margins of the Good Earth: the South Australian wheat frontier 1869–1884*. Chicago: McNally.

Mrozowski, S. 2006. Environments of history: biological dimensions of historical archaeology, in M. Hall & S. Silliman (ed.) *Historical archaeology*: 23–41. Oxford: Blackwell.

Mrozowski, S. 2010. New and forgotten paradigms: the environment and economics in historical archaeology. *Historical Archaeology* 44(3): 117–27.

Nathan, E. 2007. *Lost waters: a history of a troubled catchment*. Melbourne: Melbourne University Press.

Parliament of Victoria. 1887. Board appointed to enquire into the sludge question. Report together with the proceedings, minutes of evidence and appendices. *Victorian Parliamentary Papers*, Volume 2(10). Melbourne: Government Printer.

Phillips, G.N. & M.J. Hughes. 1998. Victorian gold deposits. *Journal of Australian Geology and Geophysics* 17(4): 213–16.

Powell, J. 1970. *The public lands of Australia Felix: settlement and land appraisal in Victoria 1834–91 with special reference to the western plains*. Melbourne: Oxford University Press.

Powell, J. 1989. *Watering the Garden State: water, land and community in Victoria 1834–1988*. Sydney: Allen and Unwin.

Ritchie, N. & R. Hooker. 1997. An archaeologist's guide to mining terminology. *Australasian Historical Archaeology* 15: 3–29.

Robin, L. 2007. *How a continent created a nation*. Sydney: UNSW Press.

Rockman, M. 2003. Knowledge and learning in the archaeology of colonization, in M. Rockman & J. Steele (ed.) *Colonization of unfamiliar landscapes: the archaeology of adaptation*: 3–24. London: Routledge.

Rockman, M. 2010. New world with a new sky: climate variability, environmental expectations, and the historical period. *Historical Archaeology* 44(3): 4–20.

Russell, G. 2005. Liquid gold: Bendigo's water supply, in M. Butcher & Y.M.J. Collins (ed.) *Bendigo at work: an industrial history*: 115–24. Strathdale (Vic.): National Trust of Australia.

Russell, G. 2009. *Water for gold! The fight to quench central Victoria's goldfields*. Melbourne: Australian Scholarly Publishing.

Serle, G. 1963. *The golden age: a history of the Colony of Victoria, 1851–1861*. Melbourne: Melbourne University Press.

Sherratt, T., T. Griffiths & L. Robin (ed.). 2005. *A change in the weather: climate and culture in Australia*. Canberra: National Museum of Australia Press.

Thorndycraft, V., D. Pirrie & A.G. Brown. 2003. An environmental approach to the archaeology of tin mining on Dartmoor, in P. Murphy & P. Wiltshire (ed.) *The environmental archaeology of industry*: 19–29. Oxford: Oxbow.

Tibby, J. 2004. Assessing the impact of early colonial Australia on the physical environment: a comment on Gale and Haworth (2002). *Archaeology in Oceania* 39(3): 144–56.

Tyrrell, I. 1999. *True gardens of the gods: Californian–Australian environmental reform, 1860–1930*. Berkeley: University of California Press.

Weston, K.S. 1992. *Minerals of Victoria* (Geological Survey Report 92). Melbourne: Department of Energy and Minerals.

Wilkinson, H.E. 1988. Bendigo geology, in D.G. Jones (ed.) *Central Victorian gold deposits*: 17–21. Perth: University of Western Australia.

Williams, M. 1974. *The making of the South Australian landscape*. London: Academic Press.

Part III
Nature and Culture

11 Exploring Human–Plant Entanglements

The Case of Australian *Dioscorea* Yams

Jennifer Atchison and Lesley Head

INTRODUCTION

Human–plant relations are fundamental to any consideration of the intersections between environment, technology and society. Plants are fundamental players in human lives, underpinning our food supply and contributing to the air we breathe, but they are easy to take for granted and have received insufficient attention in the social sciences. In the now well-established critique of the nature/culture dualism in Western thought, animal–society relations have received much more scholarly attention than human–plant relations (Hall 2011; Head et al. 2012). Within archaeology, the less durable nature of plant remains, by comparison with stone and bone, remind us of the connected invisibilities that remain under-researched: women's activities, the lives of children, patterns of transience and mobility.

Certainly the role of roots and tubers as food resources has been part of many key debates in prehistory, including the significance of starchy staples to early humans and the emergence of agriculture (Coursey 1967; Keeley 1995; O'Connell et al. 1999; Bird & O'Connell 2006). Yams and other starchy plants have been argued to be part of the repertoire of the original inhabitants of Greater Australia (Hallam 1989; Fullagar et al. 2006), with evidence now emerging from 49,000 years ago in highland New Guinea (Summerhayes et al. 2010). When archaeological and anthropological debates about hunting, gathering and agriculture turn to Australian Aboriginal environmental relations, a variety of plants with underground storage organs (USOs) emerge quickly in the literature. Examples include the *Dioscorea* yams of northern and western Australia, *Vigna lanceolata* and *Ipomoea costata* from central Australia, and *Microseris scapigera* from the southeast (Jones 1973; O'Connell & Hawkes 1981; Gott 1983; Yen 1989; Keen 2006).

While plants with USOs share a range of common features relevant to hunter–gatherer lives, there is also considerable geographic, ecological and social variability recorded in ethnographic and biogeographic literatures. Our aim in this chapter is to re-examine hunter–gatherer relations with Australian *Dioscorea* yams, with attention to both variability and patterning in their use. Although there have been over 200 years of ethnographic observation of Australian Aboriginal *Dioscorea* yam use, published in more than 60 works, there has not been a systematic overview of this literature until recently (Atchison & Head 2013). As Hildebrand has argued, "the social and environmental factors that favour or impede manipulations of yams are highly specific and nuanced" (Hildebrand 2003: 358). The particular aspect we focus on here is the assumption that starch is in itself an explanation for why Aboriginal people would invest considerable effort in yam gathering. Closer examination of the available sources shows diverse and complex interactions between plants, people, rocks, soils and digging. We argue that while starch is certainly important in connecting people and yams, the patterns of care and connection go much further. Following Meehan's (1977) argument in relation to shellfish gathering, thinking of yams in terms of "calories alone" leaves much interesting practice and variability unexamined.

We are not, however, interested in making a simple argument that 'social' factors are as important as 'ecological' ones in understanding human–yam relations. Rather, we draw on relational perspectives within the social sciences that have dismantled clear boundaries between the phenomena called 'environment,' 'society' and 'technology,' demonstrating how these concepts are constituted and assembled through different practices and materialities (Latour 1993; Whatmore 2002; Haraway 2008). Ethnographic work on plant-based practices has illustrated great diversity (Fairbairn 2005), providing important challenges to categories of human–plant relations, for example 'gatherers,' 'agriculture' and 'domestication' (Denham 2004; Fairbairn 2005). Bundles of different practices can cluster differentially in space and time into packages that we might then recognise as these bigger categories (Denham et al. 2009).

A focus on practice alone, however, gives all the agency to the humans. A relational perspective requires us to go further and consider how the plants themselves, and other variables such as the nature of the soil substrate, exert agency in this process (Head et al. 2012). We find Hodder's (2011) notion of 'entanglements' useful here. Entanglements help us think more fully about the networks into which people and yams are drawn, together and with other things, thinking in terms of associations rather than separations. They might, for example, help explain why people make particular choices, such as persisting with greater harvesting effort rather than processing effort, if harvesting also provides diverse opportunities for people to visit and reconnect with country (Head et al. 2002: 189). Our theoretical perspective in the case of yams is to step back to start with the materiality of the plants themselves, and to ask what kind of relations with humans this materiality makes possible or not possible.

In becoming archaeological remains, that is, material evidence of past entanglements, plants exert a different type of agency. Fairbairn (2005: 491) has argued that a practice-based, bottom-up focus "suggests the analytical primacy of archaeobotanical data," calling for a massive increase in research effort and application of new techniques. Preservation of material plant remains provides us some 'windows' into the past: for example, the fruit window we have previously discussed (Atchison et al. 2005). Stahl (2010) refers to these as "material moments in time." We argue it is also important to pay attention to absence, and to 'look for' plants in different ways, through the impressions, traces or outlines of their entanglements. In the case of yams, this may include discrepancies between biogeographic and ethnographic observations, and mounds of piled up stones and holes as evidence of gathering activities.

The term 'yam' has been used to describe economically useful plants that store starch in a potato-like root, tuber or rhizome. Here we follow Coursey's (1967) definition of yams as monocotyledonous angiosperms of the genus *Dioscorea*, family Dioscoreaceae from the order Dioscoreales, concentrating on the three species *D. transversa*, *D. bulbifera* and *D. hastifolia*.

With reference to our wider ethnographic review (Atchison & Head 2013), Head and Fullagar's long-term research project on the cultural landscapes of the Keep River region of the eastern Kimberley (Head & Fullagar 1997) and Atchison's PhD dissertation (Atchison 2000), we first examine how the materiality of yams enrolls humans in patterns of care. Second, we focus on the practices of yam digging, collection and processing in the context of daily life. Finally, we consider commonalities and variability in the likely trajectories of human–yam entanglement. Our research from the Keep River region was carried out during the late 1980s and through the 1990s in collaboration with senior women Biddy Simon and Polly Wandanga from the Marralam community.

THE MATERIALITY OF YAMS—ABOVE AND UNDER THE GROUND

The biogeographic perspective we detailed in Atchison and Head (2013) provides an important spatial overview at the continental scale, but it is a surficial view and is necessarily silent on what is happening beneath the ground. Botanically, the family Dioscoreaceae are shrubs or herbs that usually grow as slender twining vines (Wheeler et al. 1992). The genus *Dioscorea* is subdivided into two morphological specialisations, either rhizomatous or tuberous, and the economically useful species are all tuberous (Coursey 1967). The development and formation of tubers is probably an evolutionary adaptation to the wet–dry seasonal cycle of the monsoon environments in which most species are found. In the monsoonal savannas, *Dioscorea* have distinct periods of growth and dormancy, the stems and leaves drying up and dying off during the dry season and resprouting from the tuber during the wet (Figure 11.1). This is significant in the monsoonal north of Australia because it is only during the dry season that people finally gain access to sparsely distributed rainforest patches as the country dries out. It also coincides with the time of year when yam vines are most difficult to locate.

The yam tuber is a major store of carbohydrate, primarily as starch, which may account for 16–24% of its total weight (Hoover 2001). The proportion of total carbohydrate in an unidentified Australian *Dioscorea* sp. may be as high as 37% (Miller et al. 1993). There is little detailed biochemical analysis available for Australian yams specifically, or indeed *Dioscorea* more widely, but Hoover (2001) suggests that yam starches are significantly different from other cereal starches in structure and digestibility. The starch content itself changes throughout the dormancy period, declining with extended periods of storage and being replaced by sugars such as glucose as the tuber prepares to resprout again (Hariprakash & Nambisan 1996). In addition, *Dioscorea* tubers also contain quantities of other significant compounds found in small quantities but with important dietary activity, including storage proteins (Shewry 2003), vitamin C (Wanasundera & Ravindran 1994) and allantonin (Fu et al. 2006). Variable amounts of these constituents are also noted for Australian species (Miller et al. 1993), these studies all suggesting a range of important roles for yams in the diet other than, or in addition to, their starch content.

Most *Dioscorea* have at least two strategies of reproduction. At the onset of the wet season vines resprout from the subterranean tubers, reproducing

Figure 11.1   Excavated *Dioscorea transversa* tuber *in situ* (Photo: L. Head)

the plant vegetatively. These vines then flower and set fruit, reproducing sexually. Flower production in *Dioscorea* is dioecious, rarely monecious (Wheeler et al. 1992), and pollination is usually carried out by insects. The seeds are borne in winged, wind-dispersed capsules (Coursey 1972). *D. transversa* reproduces via these two strategies. In addition, *D. bulbifera* produces asexual aerial bulbils, genetically identical to the parent vine but able to disperse above ground. In the western Kimberley, Aboriginal people also collected and ate these bulbils (Smith & Kalotas 1985). The ethnographic record does not identify bulbils in *D. hastifolia* (Florabase 2011). Aboriginal people understand these various reproductive strategies when they break off the relatively unpalatable stringy tops of tubers and replant them at the base of the freshly dug holes, and also when they collect aerial bulbils and plant them in their home gardens (Head et al. 2002). Replanting the growing tops of yams is a consistent practice for each of the three Australian species across their wide geographic distribution. This practice deliberately reproduces plants in known locations—women remember and locate yams dug from previous years—but it also continues vegetative reproduction of yams previously dug. Women are not just digging yams from the same place when they revisit: they are often redigging and collecting the same genetic plants.

Dioscorea vines commonly twine among the shrubs and trees with which they are associated. *D. hastifolia* is also reported as being self-supporting (Florabase 2011). Rarely, the vines and undispersed seeds signal the presence of the drying-out vine to yam diggers. More commonly, people must search out fragments or remnants of yam vines among the tangle of trees and rocks and differentiate them from other unrelated climbers. To do this, a yam digger relies on her knowledge of the shape and colour of *Dioscorea* stems: *D. bulbifera* stems are sometimes slightly ridged, *D. transversa* stems often striated (Wheeler et al. 1992). As well as signalling and identifying the possibility of tubers below, the stem also provides the important physical connection from the aerial part of the plant to the subterranean tuber, and it is vital that the digger persist in maintaining and following this connection through the digging process. If that connection is broken or lost, the digger must search for it or begin again.

If the digger has successfully followed the stem to the tuber, the process of uncovering the yam tuber can begin. The tubers themselves vary considerably in size, *D. bulbifera* commonly weighing under 500g and *D. transversa* anything up to a kilogram. Hammond reported roots of *D. hastifolia* "growing up to three feet in length with a diameter from half an inch to two inches" (Hammond 1980: 28). On rocky substrates, tubers are commonly held hidden some 1–2m beneath the surface, creviced among tree roots and rocks. This strategy helps protect the vulnerable and valuable tubers from other herbivores that might smell them out. We observed regular digging and herbivory upon tubers by kangaroos in sandy substrates, but rarely on rocky soils (Atchison 2000).

YAM DIGGING AND PROCESSING PRACTICES

Digging yams is often, but not always, a labour-intensive process. In rocky country women must move large volumes of dirt and rock and cut away thick tree roots. Wooden digging sticks were commonly used in the past (see Meagher 1974) but are replaced in many places today by steel crowbars. Small hatchets are also used, although dirt is often simply scraped out by hand. Where quantification of yield exists, the results are variable. McCarthy and McArthur (1960) provide the most specific daily food collection observations for Aboriginal people relying heavily on traditional foods. In their estimates women gathered up to 21 pounds (9.5kg) in five hours, while Jones and Meehan (1989) reported skilled women were able to dig about 2–3kg of D. *transversa* in an hour. Gorman et al. (2006) reported 0.42kg per hour, with extraction being especially difficult and laborious in rocky soil. In our own studies we have estimated anything up to 300kg of dirt and rock being removed to get to an individual yam tuber (Atchison 2000) (Figure 11.2). It is important to remember that women might also spend hours looking and digging only to retrieve very little or even nothing (McCarthy & McArthur 1960).

D. *transversa* and D. *bulbifera* (McCarthy & McArthur 1960) also grow on sandy soils and can be quite easy to retrieve. Yams from sandy soils are much easier to dig; however, in the Keep River region these yams are considered too long and skinny and are not readily sought after. This is possibly because yams in sandy areas are more heavily predated and so have not fattened up over a number of years, and/or because the dolomite substrate in rocky areas provides a richer nutrient base, growing sweeter yams: "different taste [from the yams from Baranda] he got same rock longa Milyoonga" (Biddy Simon, in Atchison 2000) (Figure 11.3). Yams from rocky rainforest patches might be harder to get at, but they have other attributes that clearly make them valuable.

There are no specific time estimates available for collection of D. *hastifolia*. It is recorded on rocky (Stormon 1977; Hammond 1980), loamy (Moore 1884) and sandy (Chauncy in Smyth 1876) substrates. As previously noted, heavy soils were probably easier to dig after rainfall (Moore 1884). Chauncy observed "both men and women sinking in loose sandy soil for an edible root čalled warran, one of the Dioscoreaceae" (Chauncy in Smyth 1876: 245–6). In contrast, Grey observed in 1841 that digging for yams was a laborious activity:

> The labour in proportion to the amount obtained, is great. To get a yam about half an inch in circumference and a foot in length, they have to dig a hole above a foot square and two feet in depth; a considerable portion of the time of the women and children is, therefore, passed in this employment. (Grey 1964: 293)

Figure 11.2   Digging for *Dioscorea transversa*, Keep River, Northern Territory. In this region *D. transversa* is commonly found in small, isolated patches of monsoonal vine thicket on isolated rocky dolomite outcrops (Photo: L. Head)

Hallam (1986) quotes Grey (1840: 124) that "the digging of these roots is always a most laborious operation, and in the early dry season almost impracticable from the hard ground."

Whatever the effort involved, it does seem clear that the scale of landscape transformation as a result of digging for *D. hastifolia* is markedly different in northern Australia and the southwest. We have argued this is largely due to differences in yam habitat and soil substrate, specifically in the case of yam grounds recorded on the alluvial terraces on the coastal plains of south-western Australia (Atchison & Head 2013). Climatic differences are also relevant, as *D. hastifolia*'s window of availability is possibly longer than that of the tropical species and access to yam grounds in

Figure 11.3   Yam tuber (*Dioscorea transversa*) dug from loose sandy soil, Keep River region, Northern Territory (Photo: J. Atchison)

the southwest is not restricted by regular seasonal flooding. There are also visible influences on soil and slope processes in northern yam habitats, particularly within monsoonal vine thicket patches on rocky slopes. Tonnes of rock and soil are moved about through the tuber excavation process, which also interacts with the process of quarrying for stone artefact sources (Head et al. 2002). Nevertheless, these impacts are localised and patchy, affecting areas as small as tens of metres square (Figure 11.4). The smaller northern patches are not on the agricultural scale of yam or 'warran' grounds argued for by Hallam (1986, 1989).

Digging for yams unfolds via diverse strategies of locating and retrieval (Atchison 2000). Polly usually found a stem quickly and persevered with her choice for the duration of the visit. She was somewhat more successful

Figure 11.4   Recently excavated yam hole on small rocky dolerite outcrop, Keep River region, Northern Territory (Photo: J. Atchison)

at retrieving a yam, although often it was broken up into many small pieces. Biddy's general strategy was to locate a number of yam stems and choose the one she thought would be the easiest to get out. While she was not as consistent at retrieval as Polly, her greater strength meant she usually got yams out in a single piece. Larger pieces are considered a better shape for roasting. Over repeated days of digging for yams, it was evident that the exercise was not just about getting yams, but also a time in which the women might sit quietly in the shade and think or talk (Atchison 2000). Accompanying children might fall asleep or play as they waited. Observers were also enrolled, watching and learning, and since a digger usually kept her yams, were eventually expected to find yams for themselves (Atchison

2000). There is a necessary attentive watchfulness and quietening as the yam gradually reveals itself and unfolds other processes of daily life (Povinelli 1993).

The practices and labour of processing yams have been written about extensively in different parts of the world and a review of the Australian literature is detailed in Atchison and Head (2013). *D. transversa* is commonly eaten without any preparation, or after light roasting. Most accounts of this species are that it is sweet and very pleasant tasting and juicy. In the Keep region, *D. transversa* was the preferred species, highly sought after and readily fed uncooked to hungry and impatient children. *D. hastifolia* was also eaten raw, though more commonly reported as being roasted, or roasted and then pounded. *D. bulbifera,* on the other hand, requires extensive processing to leach out the bitter compounds (Telford 1986) before it can be eaten.

HUMAN–YAM ENTANGLEMENTS

Yam plants, the soil substrate and the rocks among which they grow are folded together into particular relations with the people who collect the yams. Collecting yams requires people to be attentive, watchful and patient. Such skills must be learnt over time and carried out within the processes of daily life, across yearly seasonal changes and even over longer cycles of renewal and regeneration. Rocks are particularly important in association with rainforest patches, contributing to the size and tastiness of yams, but also making them more difficult and laborious to extract. These rocky yam patches are small in spatial scale but are connected with other patches across a landscape of mobility. Active and regular use over the longer term seems to have been important to maintaining a supply of good quality yams. Disruption of access in the post-contact period has led to perceived variability in the health of patches, with declining yam quality both in those that are never visited and those that are over-visited. Reduced access to country and other social changes reduce the transmission of this knowledge to following generations.

It is a heavily gendered entanglement, as seen by the fact that knowledge and skill in yam digging contributes to the status of many older Aboriginal women. The daily life of Aboriginal women includes a range of activities, such as travelling, hunting, fishing and tending to children and relatives, as well as gathering and processing plants. The time and effort expended on collecting yams needs to be examined in the context of these other activities, and the interplay is most clearly documented in McCarthy and McArthur's work (1960). Tending to small children, for example, is a major consideration in planning yam collection and processing activities. In our own research, collecting yams is most often carried out on the way to another activity at another destination, commonly fishing or hunting. In this way, the time spent collecting yams and the frequency of visits is regulated. Yam patches most frequented were located along access roads. This meant they could be over-visited, in which case the women commented that collecting yams too frequently would make people sick. In contrast, other yam patches, which are remembered from the past but are not frequently visited now, do not yield yams today, possibly because the beneficial actions of people have been removed (Head et al. 2002).

In northern Australia the strong seasonality of tuber availability, in combination with its patchy distribution and sometimes tenuous visibility, means that the traditional ecological knowledge surrounding collection and retrieval is very significant in maintaining yams as part of the food supply. The specific knowledge of how to process *D. bulbifera* and make it edible adds to this significance. The investment of effort to acquire yams seems considerable and maybe inefficient, but it needs to be understood in terms of its gendered nature: other activities and broader knowledge about the landscape, parallel activities and other rituals of stewardship are facilitated during women's yam collecting activities.

Efficiency of yams as a food source, relative to the calorific effort required to dig them, seems to vary considerably with the rockiness or otherwise of the substrate. Pending more systematic research comparisons, yams should not be assumed to be easy calories just because they are starchy. This is particularly the case when the processing costs of *D. bulbifera* are considered. A variety of other reasons come into play in accounting for the favoured status of yams as food. These include sweet taste, the status and identity of yam diggers, other connections to yam places, and a lack of other food choices late in the dry season. However human use of yams started, by the time ethnographic observers were making records in Australia, people and yams had become entangled in much more complex ways than can be explained simply by starch.

ACKNOWLEDGMENTS

Biddy Simon and Polly Wandanga taught us much about yams. Andrew Warren assisted with the literature search. Tim Denham and Richard Fullagar have discussed many plant-related issues.

REFERENCES

Atchison, J. 2000. Continuity and change: a late Holocene and post contact history of Aboriginal environmental interaction and vegetation process from the Keep River region, Northern Territory. Unpublished PhD dissertation, University of Wollongong.

Atchison, J. & L. Head. 2013. Yam landscapes; the biogeography and social life of Australian *Dioscorea*, in L. Russell & Z. Ma Rhea (ed.) *The world of plants in Aboriginal Australia: essays in honour of Beth Gott. The Artefact* 35 (in press)

Atchison, J., L. Head & R. Fullagar. 2005. Archaeobotany of fruit seed processing in a monsoon savanna environment: evidence from the Keep River region, Northern Territory, Australia. *Journal of Archaeological Science* 32(2): 167–81.

Bird, D.W. & J.F. O'Connell. 2006. Behavioural ecology and archaeology. *Journal of Archaeological Research* 14(2): 143–88.

Coursey, D.G. 1967. *Yams: an account of the nature, origins, cultivation and utilisation of the useful members of the Dioscoreaceae.* London: Longmans.

Coursey, D.G. 1972. The civilizations of the yam: interrelationships of man and yams in Africa and the Indo-Pacific region. *Archaeology and Physical Anthropology in Oceania* 7(3): 215–33.

Denham, T. 2004. The roots of agriculture and aboriculture in New Guinea: looking beyond Austronesian expansion, Neolithic packages and indigenous origins. *World Archaeology* 36(4): 610–20.

Denham, T., R. Fullagar & L. Head. 2009. Plant exploitation on Sahul: from colonisation to the emergence of regional specialisation during the Holocene. *Quaternary International* 202(1–2): 29–40.

Fairbairn, A. 2005. An archaeobotanical perspective on Holocene plant-use practices in lowland northern New Guinea. *World Archaeology* 37(4): 487–502.

Florabase. 2011. Western Australian online herbarium. Department of Environment and Conservation. Available at: http://florabase.calm.wa.gov.au/ (accessed 29 March 2011).

Fu, Y., L.A. Ferng, & P. Huang. 2006. Quantitative analysis of allantoin and allantoic acid in yam tuber, mucilage, skin and bulbil of the *Dioscorea* species. *Food Chemistry* 94(4): 541–49.

Fullagar, R., J. Field, T. Denham & C. Lentfer. 2006. Early- and mid-Holocene tool-use and processing of taro (*Colocasia esculenta*), yam (*Dioscorea* sp.) and other plants at Kuk Swamp in the highlands of Papua New Guinea. *Journal of Archaeological Science* 33(5): 595–614.

Gorman, J., G. Wightman, P.J. Whitehead & J. Altman. 2006. Case study: long yam. Feasibility of small scale commercial native plant harvests by Indigenous communities, in P.J. Whitehead, J. Gorman, A.D. Griffiths, G. Wightman, H. Massarella & J. Altman (ed.) *Rural Industries Research and Development Corporation (RIRDC) Land & Water Australia Joint Venture Agroforestry Program*: 143–50. Barton, ACT: Rural Industries Research and Development Corporation.

Gott, B. 1983. *Microseris scapigera*: a study of a staple food of Victorian Aborigines. *Australian Aboriginal Studies* 2: 2–18.

Grey, G. 1840. *Vocabulary of the dialects of south-western Australia.* London: T. & W. Boone.

Grey, G. 1964. *Journals of two expeditions of discovery in north-west and western Australia, during the years 1837–1839.* London: T & W Boone.

Hall, M. 2011. *Plants as persons. A philosophical botany.* Albany: State University of New York Press.

Hallam, S.J. 1986. Yams, alluvium and 'villages' on the west coastal plain, in G.K. Ward (ed.) *Archaeology at ANZAAS Canberra*: 116–32. Canberra: Canberra Archaeological Society.

Hallam, S.J. 1989. Plant usage and management in southwest Australian Aboriginal societies, in D.R. Harris & G.C. Hillman (ed.) *Foraging and farming: the evolution of plant exploitation*: 136–51. London: Unwin Hyman.

Hammond, J.E. 1980. The camp, food and clothing, in P. Hasluck (ed.) *Winjan's People. The story of the south-west Australian Aborigines*: 25–32. Victoria Park: Hesperian Press.

Haraway, D. 2008. *When species meet.* Minneapolis: University of Minnesota Press.

Hariprakash C.S. & B. Nambisan. 1996. Carbohydrate metabolism during dormancy and sprouting in yam (*Dioscorea*) tubers: changes in carbohydrate constituents in yam (*Dioscorea*) tubers during dormancy and sprouting. *Journal of Agricultural Food Chemistry* 44(10): 3066–69.

Head, L., J. Atchison & R. Fullagar. 2002. Country and garden: ethnobotany, archaeobotany and Aboriginal landscapes near the Keep River, northwestern Australia. *Journal of Social Archaeology* 2(2): 173–96.

Head, L., J. Atchison & A. Gates. 2012. *Ingrained: a human bio-geography of wheat.* Aldershot: Ashgate.

Head, L.M. & R.L.K. Fullagar. 1997. Hunter-gatherer archaeology and pastoral contact: perspectives from the northwest Northern Territory, Australia. *World Archaeology* 28(3): 418–28

Hildebrand, E.A. 2003. Motives and opportunities for domestication: an ethnoarchaeological study in southwest Ethiopia. *Journal of Anthropological Archaeology* 22: 358–75.

Hodder, I. 2011. Human–thing entanglement: towards an integrated archaeological perspective. *Journal of the Royal Anthropological Institute* 17: 154–77.

Hoover, R. 2001. Composition, molecular structure, and physicochemical properties of tuber and root starches: a review. *Carbohydrate Polymers* 45: 253–67.

Jones, R. 1973. The Neolithic, Palaeolithic and the hunting gardeners: man and land in the Antipodes, in R.P. Suggate & M.M. Cresswell (ed.) *Proceedings of the 9th INQUA Congress*: 21–34. Wellington: The Royal Society of New Zealand.

Jones, R. & B. Meehan. 1989. Plant foods of the Gidjingali: ethnographic and archaeological perspectives from northern Australia on tuber and seed exploitation, in D.R. Harris & G.C. Hillman (ed.) *Foraging and farming: the evolution of plant exploitation*: 120–51. London: Unwin Hyman.

Keeley, L.H. 1995. Protoagricultural practices among hunter-gatherers: a cross-cultural survey, in T.D. Price & A.B. Gebauer (ed.) *Last hunters, first farmers: new perspectives on the prehistoric transition to agriculture*: 243–72. Santa Fe: School of American Research Press.

Keen, I. 2006. Constraints on the development of enduring inequalities in Late Holocene Australia. *Current Anthropology* 47: 7–38.

Latour, B. 1993. *We have never been modern.* New York: Harvester Wheatsheaf.

Mccarthy, F.D. & M. McArthur. 1960. The food quest and the time factor in Aboriginal economic life, in C.P. Mountford (ed.) *Records of the American–Australian Scientific Expedition to Arnhem Land*: 145–94. Melbourne: Melbourne University Press.

Meagher, S.J. 1974. The food resources of the Aborigines of the south-west of Western Australia. *Records of the Western Australian Museum* 3(1): 14–65.

Meehan, B. 1977. Man does not live by calories alone: the role of shellfish in a coastal cuisine, in J. Allen, J. Golson & R. Jones (ed.) *Sunda and Sahul*: 493–531. New York: Academic Press.

Miller, J.B., K.W. James & P.M.A. Maggiore. 1993. *Table of composition of Australian Aboriginal foods.* Canberra: Aboriginal Studies Press.

Moore, G.F. 1884. *A descriptive vocabulary of the language in common use amongst the Aborigines of Western Australia.* London: William S. Orr.

O'connell, J.F. & K. Hawkes. 1981. Alyawara plant use and optimal foraging theory, in B. Winterhalder & E.A. Smith (ed.) *Hunter-gatherer foraging strategies. Ethnographic and archeological analyses*: 99–125. Chicago: University of Chicago Press.

O'Connell, J.F., K. Hawkes & N.G. Blurton Jones. 1999. Grandmothering and the evolution of Homo erectus. *Journal of Human Evolution* 36: 461–85.

180   *Jennifer Atchison and Lesley Head*

Povinelli, E. 1993. 'Might be something': the language of indeterminacy in Australian Aboriginal land use. *Man* 28(4): 679–704.
Shewry, P.R. 2003. Tuber storage proteins, invited review. *Annals of Botany* 91: 755–69.
Smith, M. & A.C. Kalotas. 1985. Bardi plants: an annotated list of plants and their use by the Bardi Aborigines of Dampierland, in north-western Australia. *Records of the West Australian Museum* 12(3): 317–59.
Smyth, B.R. 1876. *The Aborigines of Victoria: with notes relating to the habits of the natives and other parts of Australia and Tasmania compiled from various sources for the Government of Victoria.* Melbourne: John Currey (reprinted 1972).
Stahl, A. 2010. Material histories, in D. Hicks & M. Beaudry (ed.) *The Oxford handbook of material culture studies*: 150–72. Oxford: Oxford University Press.
Stormon, E.J. (ed. and trans.) 1977. *The Salvado Memoirs: historical memoirs of Australia and particularly of the Benedictine mission of New Norcia and of the habits and customs of the Australian natives by Don Rosendo Salvado, O.S.B.* Nedlands: University of Western Australia Press.
Summerhayes, G.R., M. Leavesley, A. Fairbairn, H. Mandui, J. Field, A. Ford & R. Fullagar. 2010. Human adaptation and plant use in highland New Guinea 49,000 to 44,000 years ago. *Science* 330: 78–81.
Telford, I.R.H. 1986. Dioscoreaceae. *Flora of Australia.* 46: 196–219. Canberra: Australian Government Publishing Service.
Wanasundera J.P.D. & G. Ravindran. 1994. Nutritional assessment of yam (*Dioscorea alata*) tubers. *Plant Foods for Human Nutrition* 46(1): 33–9.
Whatmore, S. 2002. *Hybrid geographies. Natures cultures spaces.* London: Sage.
Wheeler, J.R., B.L. Rye, B.L. Koch & A.J.G. Wilson (ed.). 1992. *Flora of the Kimberley region.* Como WA: Western Australian Herbarium and Department of Conservation and Land Management.
Yen, D.E. 1989. The domestication of environment, in D.R. Harris & G.C. Hillman (ed.) *Foraging and farming: the evolution of plant exploitation*: 56–76. London: Unwin Hyman.

12 People and Their Environments

Do Cultural and Natural Values Intersect in the Cultural Landscapes on the World Heritage List?

Anita Smith

INTRODUCTION

In 1992 the World Heritage Committee introduced 'cultural landscape' as a new category of World Heritage cultural site in an attempt to reconnect culture and nature within the context of the 1972 World Heritage Convention. The committee was responding to the near absence of World Heritage properties reflecting the diverse and complex relationships between humans and their environments, and in particular those that characterise traditional and indigenous cultures (Fowler 2002: 19). The separation of nature from culture in the processes for inscription of properties on the World Heritage List had been identified as limiting recognition of non-Western perceptions of landscape, relationships to animals, plants, landforms and the sea, and the roles of traditional owners and custodians (Titchen 1996).

Cultural landscapes were defined by the committee as the "combined works of nature and man [*sic*]" (UNESCO 2011: para. 47), their outstanding universal values reflecting the interaction of humans and the environment. While providing a pathway for increasing the representation of places significant to traditional and indigenous peoples on the World Heritage List, introduction of the cultural landscape category was also anticipated to increase the representation of regions such as Africa, the Pacific, and Latin America and the Caribbean on the World Heritage List, as well as the diversity of site types and values. In 2012, 10% of properties inscribed on the World Heritage List are cultural landscapes, a figure often cited as evidence of the opening of the World Heritage Convention to cultures in regions other than Europe that were not represented or were under-represented prior to 1992, a recognition of the nonmonumental character of the heritage of cultural landscapes and acknowledgement of the links between cultural and biological diversity (Bandarin 2007: 3).

While the popularity of cultural landscapes in the World Heritage system is clear, any claims for the success of the category as a vehicle for better recognising the interaction of people and the environment and, in particular, the heritage of traditional cultures and indigenous peoples warrants closer scrutiny. Review of the cultural landscapes inscribed on the World Heritage

List indicates that while the cultural landscape category may have filled a significant gap in the World Heritage List, only a handful of these properties reflect the values of indigenous and traditional peoples from those regions least represented on the World Heritage List.

BACKGROUND

The UNESCO Convention Concerning the Protection of the World Cultural and Natural Heritage 1972, better known as the World Heritage Convention 1972, is the only international instrument that explicitly defines and protects both cultural and natural heritage. The aim was to create an "effective system of collective protection of the cultural and natural heritage of outstanding universal value" (UNESCO 1972), in particular through the creation of a World Heritage List. However, the processes and procedures for inscription of properties on the list and systems for their protection and management perpetuated a distinction between cultural and natural heritage values and universalised the Western scientific construct of 'nature' as distinct from culture (Plachter & Rossler 1995: 16). All World Heritage properties must be demonstrated to have "outstanding universal value" (UNESCO 2011: para. 49–53) and are inscribed on the World Heritage List as cultural or natural properties, satisfying at least one of six cultural criteria (Criteria i–vi) or four natural criteria (Criteria vii–x), respectively. Mixed properties, with both cultural and natural values that are considered outstanding, satisfy at least one cultural and one natural criterion.

A rigid distinction between the conceptual frameworks in which cultural and natural values were articulated and evaluated was apparent from the initial inscriptions of World Heritage properties in 1978. Natural sites were valued as pristine or wilderness not compromised by human use or occupation, while the inscriptions of cultural properties were dominated by the unique and grand in architecture, monuments and historic towns reinforcing a dualistic, static and conservative approach to protection (Plachter & Rossler 1995: 16). By the late 1980s it had become clear that this conservative and essentially Eurocentric approach was contributing to a growing imbalance in the representation of cultural sites and geo-cultural regions on the World Heritage List (UNESCO 1994a). The architectural heritage of Europe dominated the list, while the cultural heritage of Africa, the Pacific Islands, and Latin America and the Caribbean was significantly underrepresented, a situation that was widely perceived to threaten the credibility of the list as a record of the outstanding heritage of humanity (UNESCO 1994a).

The under-representation of these regions could in part be explained by the lack of capacity for heritage protection in many developing countries and, therefore, a lack of capacity for developing successful World Heritage nominations. However, the Operational Guidelines to the World Heritage

Convention defined a limited range of cultural site types that could be nominated, and these did not include landscapes (UNESCO 1992: para. 23–30). As early as the mid-1980s this had been seen as limiting the diversity of European sites that could be considered for nomination (UNESCO 1987) and by the late 1980s as contributing to the growing regional imbalance of the list.

Recognition of the need to broaden the understanding of heritage values within the World Heritage system reflected wider contemporary debates in archaeology (Gosden & Head 1994) and geography (Aplin 2007: 428–9) around the significance of landscape as both a conceptual and practical framework for understanding human behaviour in the past and present (Lowenthal 1997). Alongside this, the increasing participation of indigenous and other traditional communities in heritage management prompted recognition of the need for culturally appropriate approaches to the interpretation and management of their heritage. This was mirrored in natural heritage conservation by increasing awareness of the role of communities, especially indigenous communities, in the creation and management of 'natural' landscapes (Bridgewater et al. 2007).

CULTURAL LANDSCAPES ON THE WORLD HERITAGE LIST

The introduction of the cultural landscape category in 1992 meant that for the first time within the history of the convention landscapes were acknowledged to be the outcome of cultural processes and nominated as *cultural* properties:

> . . . illustrative of the evolution of human society and settlement over time, under the influence of the physical constraints and/or opportunities presented by their natural environment and of successive social, economic, and cultural forces, both external and internal. (UNESCO 2011: para. 47)

Prior to 1992 landscapes in which human interaction with the environment contributed to the outstanding universal values could be nominated as natural sites, on what was then Natural Criterion ii ("outstanding examples representing significant ongoing geological processes, biological evolution and man's interaction with his natural environment") or Natural Criterion iii ("contain superlative natural phenomena, formations or features . . . or exceptional combinations of natural and cultural elements") (UNESCO 1992: para. 36). Although these natural criteria were frequently used, none of the then 82 natural properties on the World Heritage List was considered to have outstanding universal value on the basis of human interaction with the environment, and only a handful made reference to the significance of cultural 'elements.' Prior to 1992 landscapes could also be nominated as mixed properties on both cultural and natural criteria. In almost all the 16 mixed properties inscribed prior to the introduction of the cultural land-scape category, the cultural values are evidenced in archaeological and/or rock art sites, for example, Kakadu National Park in Australia. In nomina-tions of mixed properties the cultural values and associated evidence are presented and evaluated in isolation from the natural values and are seen as co-existent rather than interrelated, a concept Fowler (2002: 17) argues was 'intellectually flaccid' in not providing a framework to move beyond a combination of features to recognise interplay between cultural and natural influences.

With the introduction of cultural landscape as a cultural site type, ref-erences to 'human interaction with the environment' and 'cultural ele-ments' were removed from Natural Criteria ii and iii (now Criteria ix and vii, respectively) and cultural Criterion v was revised, becoming 'to be an outstanding example of traditional human settlement or land-use which is representative of a culture (or cultures) especially when it has become vulner-able under the impact of irreversible change' (UNESCO 1994b: para. 24). Also of particular relevance for cultural landscapes where traditional asso-ciations or stories frame understanding of the landscape is Criterion vi: 'to be directly or tangibly associated with events or living traditions, with ideas, or with beliefs, with artistic and literary works of outstanding universal significance."

To assist in the nomination and evaluation of potentially very diverse cul-tural landscapes the World Heritage Committee also defined three catego-ries of cultural landscapes (UNESCO 2011: annex 3, para. 10) (Table 12.1):

- Category i: a clearly defined landscape designed and created intention-ally by man [*sic*]
- Category ii: an organically evolved landscape
- Category iii: an associative cultural landscape.

In 2012 there are 936 properties on the World Heritage List, of which 93 are cultural landscapes. Details of the cultural landscapes inscribed on the World Heritage List are shown in Table 12.2

Most World Heritage cultural landscapes are Category ii 'organically evolved' landscapes, that is, the cultural evidence reflecting the interactions of humans and the environment has evolved over time. A majority of these (44) are Category iib, "continuing cultural landscapes," in which the cul-tural and social processes that created the landscape are continuing. Thirty are Category iia, "relic" landscapes, where the cultural processes that cre-ated the landscape came to an end at some time in the past. There are ten Category i, clearly defined, intentionally created landscapes, nine of which are in Europe. There have been only nine inscriptions of cultural landscapes in which the primary values are intangible or associative.

Table 12.1   World Heritage Cultural Landscape Categories (UNESCO 2011: annex 3, para. 10)

---

i.   **The clearly defined landscape designed and created intentionally by man.** This embraces garden and parkland landscapes constructed for aesthetic reasons which are often (but not always) associated with religious or other monumental buildings and ensembles.

ii. **The organically evolved landscape.** This results from an initial social, economic, administrative, and/or religious imperative and has developed its present form by association with and in response to its natural environment. Such landscapes reflect that process of evolution in their form and component features. They fall into two sub-categories:
    a. a relict (or fossil) landscape is one in which an evolutionary process came to an end at some time in the past, either abruptly or over a period. Its significant distinguishing features are, however, still visible in material form.
    b. a continuing landscape is one which retains an active social role in contemporary society closely associated with the traditional way of life, and in which the evolutionary process is still in progress. At the same time it exhibits significant material evidence of its evolution over time.

iii. **The associative cultural landscape.** The inscription of such landscapes on the World Heritage List is justifiable by virtue of the powerful religious, artistic or cultural associations of the natural element rather than material cultural evidence, which may be insignificant or even absent.

Fifty cultural landscapes are located in Europe and North America, 22 in the Asia–Pacific region (two of which are in independent Pacific Island states), 12 in African countries, 7 in Latin America and the Caribbean, and 2 in the Arab states. The disproportionate regional representation of these properties is a pattern seen in World Heritage inscriptions in general, almost 50% of which are located in Europe and North America, with only 8% in Africa.

In Europe the "continuing cultural landscapes' with outstanding universal values, 27 in total, are associated with the origin, evolution and persistence of settlements and/or land use in a specific geographic zone or ecosystem. These properties provide two distinct forms of evidence. The contemporary landscape is *either* characterised by traditional land use practices—agricultural (the Alto Douro Wine Region, Portugal) or pastoral (the Causses and the Cévennes, France) and associated settlements—*or* is a palimpsest of evidence, built and/or archaeological, from successive cultural phases in a specific landform or feature (e.g. the Upper Middle Rhine Valley in Germany and the coastal zone of Costiera Amalfitana in Italy). Relic cultural landscapes are few in number. They include mining and industrial sites such as the Derwent Valley Mills (UK) and the Cornwall and West Devon Mining Landscape (UK), respectively, and the archaeological remains of ancient settlements, such as Kernavė Archaeological Site (Lithuania). The nine

Table 12.2 Cultural landscapes inscribed on the World Heritage List in 2012

UNESCO region	Name of property	Country	Year inscribed	Site type	Landscape category (iia, iib, iii)	World Heritage Criteria (i–x)	Theme/Values
AFR	Sukur Cultural Landscape	NG	1999	C	iib, iii	iii, v, vi	traditional settlement/land use
AFR	Mapungubwe Cultural Landscape	ZA	2003	C	iia	ii, iii, iv, v	archaeological settlement/land use
AFR	Ngorongoro Conservation Area	TZ	2010 (1978)	M	iia	iv, vii, viii, ix, x	human evolution, prehistoric
AFR	Ecosystem and Relict Cultural Landscape of Lopé-Okanda	GA	2006	M	iia	iii, iv, ix, x	Successive cultural phases, valley (archaeological)
AFR	Matobo Hills	ZW	2003	C	iia, iib	iii, iv, v	rock art
AFR	Tsodilo	BW	2001	C	iib, iii	i, iii, vi	rock art
AFR	Royal Hill of Ambohimanga	MG	2001	C	iib, iii	iii, iv, vi	sacred landscape
AFR	Sacred Mijikenda Kaya Forests	KE	2008	C	iib, iii	iii, v, vi	sacred landscape
AFR	Osun-Osogbo Sacred Grove	NG	2005	C	iib, iii	ii, iii, vi	sacred landscape
AFR	The Konso Cultural Landscape	ET	2011	C	iib	i, ii, v	traditional agriculture/settlement
AFR	Richtersveld Cultural and Botanical Landscape	ZA	2007	M	iib	iv, v	traditional pastoral landscape

AFR	Koutammakou, the Land of the Batammariba	TG	2004	C	iib, iii	v, vi	traditional settlement/ land use
APA	Gusuku Sites and Related Properties of the Kingdom Ryukyu	JP	2000	C	iia, iii	ii, iii, vi	traditional settlement, associative
APA	Parthian Fortresses of Nisa	TM	2007	C	iia	ii, iii	archaeological settlement
APA	Tongariro National Park	NZ	1993 (1990)	M	iii	vi, vii, viii	associative (indigenous tradition)
APA	Chief Roi Mata's Domain	VU	2008	C	iii, iib	iii, v, vi	associative (indigenous tradition)
APA	West Lake Cultural Landscape of Hangzhou	CN	2011	C	i, iii	ii, iii, vi	designed garden landscape
APA	Uluru-Kata Tjuta National Park	AU	1994 (1987)	M	iii, iib	iv, vi, vii, viii	associative (indigenous tradition)
APA	Iwami Ginzan Silver Mine and its Cultural Landscape	JP	2007	C	iia	ii, iii, v	mining, industrial
APA	Cultural Landscape and Archaeological Remains of the Bamiyan Valley	AF	2003	C	iia, iib, iii	i, ii, iii, iv, vi	successive cultural phases, river valley
APA	Orkhon Valley Cultural Landscape	MN	2004	C	iib	ii, iii, iv	pastoral (nomadic) landscape
APA	The Kuk Early Agricultural Site	PG	2008	C	iia	iii, iv	archaeological (early agriculture)

(Continued)

Table 12.2—(Continued)

UNESCO region	Name of property	Country	Year inscribed	Site type	Landscape category (iia, iib, iii)	World Heritage Criteria (i–x)	Theme/Values
APA	Petroglyphic Complexes of the Mongolian Altai	MN	2011	C	iia	iii	rock art
APA	Rock Shelters of Bhimbetka	IN	2003	C	iib	iii, v	rock art
APA	Petroglyphs within the Archaeological Landscape of Tamgaly	KZ	2004	C	iia, iib	iii	rock art / archaeological settlements
APA	Lushan National Park	CN	1996	C	iii, iib	ii, iii, iv, vi	sacred landscape
APA	Sulaiman-Too Sacred Mountain	KG	2009	C	iii, iib	iii, vi	sacred landscape
APA	Shrines and Temples of Nikko	JP	1999	C	iib, iii	i, iv, vi	sacred landscape
APA	Sacred Sites and Pilgrimage Routes in the Kii Mountain Range	JP	2004	C	iii, iib	ii, iii, iv, vi	sacred landscape
APA	Vat Phou and Ancient Settlements of Champasak Cultural Landscape	LA	2001	C	iia, iii	iii, iv, vi	sacred landscape and settlement
APA	Mount Wutai	CN	2009	C	iii, iib	ii, iii, iv, vi	sacred landscape
APA	Rice Terraces of the Philippines Cordilleras	PH	1995	C	iib	iii, iv, v	traditional agriculture
APA	Bam and its Cultural Landscape 2004	IR	2004	C	iib	ii, iii, iv, v	traditional settlement

Region	Name	Country	Year	Type	Code	Criteria	Description
APA	Historic Monuments of Ancient Nara	JP	1998	C	iib	ii, iv	traditional settlement
ARB	Shisr etc (Oman) – Land of Frankincense	OM	2002	C	iia	iii, iv	archaeological settlement
ARB	Cultural Sites of Al Ain (Hafit, Hili, Bidaa Bint Saud and Oases Areas)	AE	2011		iia		archaeological, hydrological
EUR	Upper Middle Rhine Valley	DE	2002	C	iib	ii, iv, v	Successive cultural phases, river valley
EUR	Kernavé Archaeological Site (Cultural Reserve of Kernavé)	LT	2004	C	iia	iii, iv	archaeological settlement
EUR	Cilento & Vallo di Diano National Park, Paestum, Velia, Certosa di Padula	IT	1998	C	iia	iii, iv	archaeological settlements
EUR	Papahānaumokuākea Marine National Monument	US	2010	M	iii, iia	iii, vi, viii, ix, x	associative (indigenous tradition)
EUR	Pingvellir National Park	IS	2004	C	iii, iia	iii, vi	associative (traditional government)
EUR	Muskauer Park / Park Mużakowski	DE/PL	2004	C	i	i, iv	designed garden landscape
EUR	Royal Botanic Gardens, Kew	GB	2003	C	i	ii, iii, iv	designed garden landscape
EUR	Aranjuez Cultural Landscape	ES	2001	C	i	ii, iv	designed garden landscape

(Continued)

Table 12.2—(Continued)

UNESCO region	Name of property	Country	Year inscribed	Site type	Landscape category (iia, iib, iii)	World Heritage Criteria (i–x)	Theme/Values
EUR	Garden Kingdom of Dessau-Wörlitz	DE	2000	C	i	ii, iv	designed garden landscape
EUR	The gardens and castle at Kromeriz	CZ	1998	C	i	ii, iv	designed garden landscape
EUR	Lednice-Valtice Cultural Landscape	CZ	1997	C	i	i, ii, iv	designed garden landscape
EUR	Kalwaria Zebrzydowska: Mannerist Architectural and Park Landscape	PL	1999	C	i, iii	ii, vi	designed garden landscape
EUR	Derwent Valley Mills	GB	2001	C	iia	ii, iv	industrial
EUR	Blaenavon Industrial Landscape	GB	2001	C	iia	iii, iv	industrial
EUR	Hallstatt-Dachstein / Salzkammergut Cultural Landscape	AT	1997	C	iib	iii, iv	mining and traditional settlement
EUR	Roros Mining Town and the Circumference	NO	2010	C	iia	iii, iv, v	mining, industrial
EUR	Cornwall and West Devon Mining Landscape	GB	2006	C	iia	ii, iii, iv	mining, industrial

EUR	The historic cultural landscape of the Great Copper Mountain in Falun	SE	2001	C	iia	ii, iii, v	mining, industrial
EUR	The Cultural Industrial Landscape of the Zollverein Mine	DE	2001	C	iia	ii, iii	mining, industrial
EUR	Ibiza, Biodiversity and Culture	ES	1999	M	iib	ii, iii, iv, ix, x	successive cultural phases, coastal landscape
EUR	Portovenere, Cinque Terre, and the Islands (Palmaria, Tino and Tinetto)	IT	1997	C	iib	ii, iv, v	successive cultural phases, coastal landscape
EUR	Costiera Amalfitana	IT	1997	C	iib	ii, iv, v	successive cultural phases, coastal landscape
EUR	Agricultural Landscape of Southern Öland	SE	2000	C	iib	iv, v	successive cultural phases, plateau
EUR	The Loire Valley between Sully-sur-Loire and Chalonnes	FR	2000	C	iib	i, ii, iv	successive cultural phases, river valley
EUR	Wachau Cultural Landscape	AT	2000	C	iib	ii, iv	successive cultural phases, river valley
EUR	Droogmakerij de Beemster	NL	1999	C	i	i, ii, iv	reclaimed landscape
EUR	Dresden Elbe Valley Delisted 2009	DE	2004	C	iib	ii, iii, iv, v	successive cultural phases, river valley
EUR	Gobustan Rock Art Cultural Landscape	AZ	2007	C	iia	iii	rock art

(*Continued*)

Table 12.2—(Continued)

UNESCO region	Name of property	Country	Year inscribed	Site type	Landscape category (iia, iib, iii)	World Heritage Criteria (i–x)	Theme/Values
EUR	Sacri Monti of Piedmont and Lombardy	IT	2003	C	iib	ii, iv	sacred landscape (mountain)
EUR	Incense Route - Desert Cities in the Negev	IL	2005	C	iia	iii, v	trade route
EUR	Landscape of the Pico Island Vineyard Culture	PT	2004	C	iib	iii, v	traditional agriculture
EUR	Stari Grad Plain	HR	2008	C	iib	ii, iii, v	traditional agriculture/ settlement
EUR	Lavaux, Vineyard Terraces	ZW	2007	C	iib	iii, iv, v	traditional agriculture/ settlement
EUR	Tokaj Wine Region Historic Cultural Landscape	HU	2002	C	iib	iii, v	traditional agriculture/ settlement
EUR	Alto Douro Wine Region	PT	2001	C	iib	iii, iv, v	traditional agriculture/ settlement
EUR	Cultural Landscape of the Serra de Tramuntana 2011	ES	2000	C	iib	ii, iv, v	traditional agriculture/ settlement
EUR	Jurisdiction of Saint-Emilion	FR	1999	C	iib	iii, iv	traditional agriculture/ settlement
EUR	Val d'Orcia	IT	2004	C	iib, iii	iv, vi	traditional agriculture/ settlement
EUR	Hortobágy National Park – the Puszta	HU	1999	C	iia, iib	iv, v	traditional pastoral landscape

EUR	The Causses and the Cévennes	FR	2011	C	iib	Iii, v	traditional pastoral landscape
EUR	Madriu-Perafita-Claror Valley	AD	2004	C	iib	v	traditional pastoral landscape
EUR	Fertö / Neusiedlersee Cultural Landscape	AT	2001	C	iib	v	traditional pastoral landscape
EUR	Pyrénées – Mont Perdu	FR/SN	1997	M	iib	iii, iv, v, vii, viii	traditional pastoral landscape
EUR	Vegaøyan – The Vega Archipelago	NO	2004	C	iib	v	traditional settlement
EUR	Curonian Spit	LT	2000	C	iia, iib	v	traditional settlement/ land use
EUR	St Kilda	GB	(2005) 1986	M	iia	iii, v, vii ix, x	traditional settlement/ grazing
EUR	Cultural Landscape of Sintra	PT	1997	C	iib	ii, iv, v	traditional settlement
EUR	Laponian Area	SW	1996	M	iib	ii, v, vii, viii, ix	transhumance herding
EUR	Rhaetian Railway in the Albula / Bernina Landscapes	IT	2008	C	iib	ii, iv	transport route
EUR	The Semmering Railway	AT	1998	C	iib	ii, iv	transport route
LAC	Quebrada de Humahuaca	AR	2003	C	iib	ii, iv, v	successive cultural phases, valley

(Continued)

Table 12.2.—(Continued)

UNESCO region	Name of property	Country	Year inscribed	Site type	Landscape category (iia, iib, iii)	World Heritage Criteria (i–x)	Theme/Values
LAC	Prehistoric caves of Yagul & Mitla in the Central Valley of Oaxaca	MX	2010	C	iia	iii	archaeological (early agriculture)
LAC	Sewell Mining Town	CL	2006	C	iia	ii	mining/ industrial settlement
LAC	Agave Landscape and Ancient Industrial Facilities of Tequila	MX	2006	C	iib, iii	ii, iv, v, vi	traditional agriculture
LAC	Archaeological Landscape of First Coffee Plantations, South-East Cuba	CU	2000	C	iia	iii, iv	traditional agriculture/ settlement
LAC	Viñales Valley	CU	1999	C	iib	iv	traditional agriculture/ settlement
LAC	Coffee cultural landscape	Col	2011	C	iib, iii	v, vi	traditional agriculture/ settlement

Source: UNESCO World Heritage Centre, whc.unesco.org. Accessed 8 January 2012.

designed landscapes in Europe are all gardens or parklands constructed for aesthetic reasons and/or to embody a philosophical tradition or the prestige of an individual or family (e.g. Gardens and Castle at Kromeríz, Czech Republic). All are associated with built or monumental heritage. A single primarily associative cultural landscape in Europe is Pingvellir National Park in Iceland, the meeting place of the ancestral Icelandic chiefs.

The character of the seven cultural landscapes in Latin America and the Caribbean is similar to those in Europe. Three are continuing cultural landscapes of agricultural production (coffee, tobacco and agave plantations and their associated settlements); four are relic landscapes, including mining (Sewell Mining Town, Chile) and agricultural properties (Archaeological Landscape of First Coffee Plantations, Cuba), and the archaeological sites of prehistoric caves (Yagul and Mitla in the Central Valley of Oaxaca, Mexico, discussed below).

Both of the cultural landscapes in the Arab states are relic landscapes, archaeological sites that are testimony to ancient trade (Land of Frankincense, Oman) and hydrological systems (Cultural Sites of Al Ain, United Arab Emirates). In both, archaeological remains of extensive built structures constitute the primary indicators of the outstanding universal value of the property.

The single designed cultural landscape in the Asia–Pacific region is the West Lake Cultural Landscape of Hangzhou (China), a landscape of pagodas, gardens and lakes. Scenic beauty, architecture and design principles also characterise many continuing cultural landscapes in Asia (Natsuko & Sirisrisak 2008), including the sacred mountains and built heritage of Islamic and Buddhist traditions (Sulaiman-Too in Krgyzstan and Mt Wutai in China), settlements (Historic Monuments of Ancient Nara, Japan) and the associative sacred landscape of Lushan National Park (China). Relic landscapes in the region include mining landscapes (Iwami Ginzan Silver Mine and its Cultural Landscape, Japan) and ancient political centres (Gusuku Sites and Related Properties of the Kingdom Ryukyu, Japan). There are also several relic and continuing cultural landscapes in UNESCO's Asia–Pacific region that reflect the values of traditional and indigenous peoples. These are discussed below, together with the 12 inscriptions of cultural landscapes from African countries.

Discussion of the individual values and evidence for the inscription of each of the above cultural landscapes is beyond the scope of this chapter. As a whole they reflect a great diversity in cultural values and in their expression in the landscape. In almost all these properties the primary indicator of the outstanding universal values of the property is built and/or archaeological heritage. Where described, the environmental evidence or natural 'elements' supporting interpretation of the properties as cultural landscapes consists primarily in the association of land or resource use with an environmental zone or the physical association of built and/or archaeological heritage with a particular landform or feature. In almost all these properties environment has been significantly modified by human action and/or human land use practices. 'Interaction' between humans and their environment is tangibly expressed in large-scale modification of the landscape, especially through the construction of structures and settlements, and agricultural production. With the exception of a handful of associative cultural landscapes, in which the interrelationships between humans and the environment are explicit in the manifestation of cultural or religious traditions in 'natural' landforms, the outstanding universal values of the cultural landscapes do not describe culturally specific understandings of landscape or relationships between the people or communities creating or using these landscapes and the 'natural' world. It was the potential for the cultural landscapes category to recognise culturally specific relationships to and interaction with the environment that underpinned claims for its relevance to the heritage of traditional, non-European cultures, 'hunter–gatherers' and other indigenous peoples.

There have been only 23 successful cultural landscape nominations associated with 'hunter–gatherers' or traditional non-European societies. These properties are not associated with traditions or other evidence from the major religions or political elites of state societies of the last 2000 years. Twelve are in Africa, nine in the Asia–Pacific region (including Papahānaumokuākea National Marine Monument, Hawai'i, USA, formally within the UNESCO region of Europe and North America), and one in each of Europe and Latin America. They are listed according to the categories of cultural landscape on which they have been inscribed (Table 12.3), although the majority of these properties could be considered in more than one category.

Four of these properties are Category iia relic cultural landscapes in which the primary indicator of outstanding universal value is rock art (Petroglyphs of the Mongolian Altai, Mongolia; Petroglyphs within the Archaeological Landscape of Tamgaly, Kazakhstan; Gobustan Rock Art Cultural Landscape, Azerbaijan; Matobo Hills, Zimbabwe). Rock art is also the primary evidence in two of the eight Category iib continuing cultural landscapes (Tsodilo, Botswana and the Rock shelters of Bhimbetka, India).

Where the interaction between humans and the environment is referred to in the outstanding universal values of these properties, this is argued to be evident in the motifs of the art: in particular, depictions of animals and plants and the use of natural resources, including hunting. In these properties the extent of the cultural landscape is defined either by the geological formation on which the art is found (e.g. the small area of massive quartzite rock formations in Tsodilo, Zimbabwe) or by a boundary unrelated to the cultural values of the property (e.g. that of a pre-existing national park for Petroglyphs of the Mongolian Altai, Mongolia). Although the rock art is said to reflect a 'cultural landscape,' the extent to which the tangible evidence equates with the landscape of a cultural community is unclear and not articulated in the values of these properties. An association with a contemporary cultural community is recognised in the values of both the African cultural landscapes characterised by rock art and the

Table 12.3 World Heritage cultural landscapes associated with 'hunter–gatherers' or traditional (non-European) societies

iia Relic cultural landscapes:
Kuk Early Agricultural Site (Papua New Guinea); Prehistoric Caves of Yagul and Mitla in the Central Valley of Oaxaca (Mexico); Ecosystem and Relict Cultural Landscape of Lopé-Okanda (Gabon); Ngorongoro Conservation Area (Tanzania); Mapungubwe Cultural Landscape (South Africa); Matobo Hills (Zimbabwe); Petroglyphs of the Mongolian Altai (Mongolia); Petroglyphs within the Archaeological Landscape of Tamgaly (Kazakhstan); Gobustan Rock Art Cultural Landscape (Azerbaijan)

iib Continuing cultural landscapes:
Koutammakou, the Land of the Batammariba (Togo); Tsodilo (Botswana); the Rock shelters of Bhimbtetka (India); Sukur Cultural Landscape (Nigeria); Rice Terraces of the Philippines Cordilleras (Philippines); The Konso Cultural Landscape (Ethiopia); Richtersveld Cultural and Botanical Landscape (South Africa); Orkhon Valley (Mongolia)

iii Associative cultural landscapes:
Papahānaumokuākea Marine National Monument (USA), Osun-Osogbo Sacred Grove (Nigeria), Uluru Kata-Tjuta National Park (Australia), Tongariro National Park (Aotearoa/New Zealand), Sacred Mijikenda Kaya Forests (Kenya), Chief Roi Mata's Domain (Vanuatu), Royal Hill of Ambohimanga (Madagascar)

rock shelters of Bhimbetka, India. Despite this, it is the geographic extent of the rock art sites that defines the property rather than the landscape of the contemporary community that produces the art. The definition of the boundaries of these properties and the description of and values attributed to the rock art are not significantly different from those of rock art sites inscribed on the Heritage List as cultural 'sites,' based on the extent of the physical evidence. An example is Chongoni Rock Art Area, Malawi, in which rock art is said to record the cultural history and traditions of the peoples of the Malawi plateau and is associated with living cultural traditions.

In several of the cultural landscapes whose boundaries and values are defined by rock art, archaeological evidence for past human activity is also an indicator of outstanding universal value of the property (Petroglyphs within the Archaeological Landscape of Tamgaly, Kazakhstan; Gobustan Rock Art Cultural Landscape, Azerbaijan; Matobo, Zimbabwe). In all of these the archaeological deposits are argued to reflect continuing human use of the region by successive cultural groups over a long period of time. This is not dissimilar to the case in cultural landscapes in Europe, where a palimpsest of cultural evidence is associated with a particular landform in which the association of tangible cultural heritage with a particular landform defines the physical extent of the cultural landscape rather than the interaction between humans and the environment that may be reflected in the evidence.

Two relic cultural landscapes associated with hunter–gatherers have been inscribed on archaeological evidence alone, the Cultural Landscape of Lopé-Okanda (Gabon) and Ngorongoro Conservation Area (Tanzania). In both the evidence for early human/hominid use of the area underpins the outstanding universal values of the property. These are mixed properties, inscribed on natural and cultural criteria, and both properties were recognised and protected for their natural values as national parks prior to nomination as cultural landscape. Ngorongoro Conservation Area (Tanzania) was inscribed as a natural site in 1978 and in 2010 was successfully renominated as a relic cultural landscape. The boundaries of the national parks were retained as the boundaries of these cultural landscapes even though the extent to which these boundaries are meaningful in terms of the archaeological evidence is unclear. In these properties the archaeological evidence, although visible in 'sites' at some locations, is largely ephemeral and dispersed across the surface and subsurface deposits of the landscape, as it is likely to be in all landscapes associated with hunter–gatherers. As Mussi (2010: 190) has recently argued, there is a need to recognise that in archaeological landscapes such as these culture and nature come together not just in the human behaviour that created the archaeological material but in the creation of the landscape, a dynamic landscape. Seen from this perspective, the properties accord more closely with the idea of an evolving archaeological or cultural landscape.

No cultural landscapes on the World Heritage List recognise dynamism in the creation of archaeological landscapes in their outstanding universal values. However, the values of two relic cultural landscapes associated with the early cultivation of plants—the Prehistoric Caves of Yagul and Mitla in the Central Valley of Oaxaca (Mexico) and Kuk Early Agricultural Site (Papua New Guinea)—do interpret the archaeological deposits in relation to both the palaeoenvironmental evidence and the extant environment to support the values on which the properties are inscribed and their inscription as landscapes rather than archaeological sites. In both, significance of the archaeology is explained in relation to environment (plant) evidence.

The remainder of the continuing cultural landscapes are landscapes of traditional agriculture (Rice Terraces of the Philippines Cordilleras) in association with culturally specific settlement patterns and architecture (Konso Cultural Landscape, Ethiopia; Sukur Cultural Landscape, Nigeria; Koutammakou, the Land of the Batammariba, Togo), or the landscapes of nomadic herders and pastoralists (Richtersveld Cultural and Botanical Landscape, Orkhon Valley, Mongolia). In each property the outstanding universal values express the cultural values of a specific community. While in most the tangible evidence, like that of the European and Latin American cultural landscapes, is primarily related to land use and settlement pattern, the significance of this evidence is articulated within an anthropological framework in which the physical evidence is an indicator of economic and social systems and/or beliefs of a contemporary cultural community. The values of these properties also explicitly refer to the cultural landscape as expressed by cultural and social responses to the constraints and opportunities offered by the environment.

There are seven indigenous cultural landscapes. Four are primarily associative. Osun-Osogbo Sacred Grove (Nigeria) and Sacred Mijikenda Kaya Forests (Kenya) are 'natural' forested areas protected and cared for by the local community as sacred places. The associative values of Chief Roi Mata's Domain (Vanuatu) and the Royal Hill of Ambohimanga (Madagascar) reflect traditional systems of authority. In their celebration of indigenous intangible cultural values the description and interpretation of the patterned landscape in all four properties provide important models for the cultural landscape nominations in developing countries, although only the values of Osun-Osogbo Sacred Grove (Nigeria) and Sacred Mijikenda Kaya Forests (Kenya) explicitly connect cultural and natural evidence as their values. Of the remaining three associative cultural landscapes, Uluru Kata-Tjuta National Park (Australia) and Tongariro National Park (Aotearoa/New Zealand) were initially inscribed as natural sites in 1987 and 1990, respectively, before being reinscribed for their indigenous cultural values in 1994 and 1993, respectively. Papahānaumokuākea National Marine Monument (Hawai'i, USA), inscribed as a cultural landscape in 2010 for its traditional significance for living Native Hawaiian culture, is also a mixed site and pre-existing natural protected area. All three properties contain significant archaeological evidence but rather than human interaction with the environment being evidenced through the physical evidence, it is associative values—indigenous knowledge—that frames understanding of individual landforms and features and the landscape.

DISCUSSION

The aim of the World Heritage Committee in introducing the cultural landscape category in 1992 was threefold: to recognise the heritage values of landscapes on the World Heritage List, to provide a vehicle in the World Heritage system to recognise the outstanding universal values of human interaction with the environment and to increase the representation of non-European, traditional and indigenous cultures on the World Heritage List. The large number and diversity of cultural landscapes now inscribed on the World Heritage List confirms the significance of landscape as a record of human activity, cultural practices and social systems associated with particular environments, ecological zones or land features and demonstrates within the World Heritage system a degree of resilience and fluidity in accommodating new ways of thinking about heritage and the value of cultural heritage to different groups of people (Titchen 1996: 240). Through providing a pathway for, in particular, the inclusion of a wide range of land use practices, designed landscapes and those reflecting religious and philosophical traditions, the cultural landscape category significantly broadened the heritage values and associated evidence of World Heritage properties, contributing to a more representative, culturally diverse and egalitarian World Heritage List.

The success of the cultural landscape category in bringing culture and nature together in the World Heritage system is, however, less clear. Cultural landscapes as a category of World Heritage property, and more specifically the three types of landscape (designed, organically evolved and associative), 'reconnect' culture and nature within the World Heritage List by recognising the values of human interaction with the landscape through two distinct forms of evidence:

- tangible evidence, that is, the physical evidence of the landscape argued to reflect the interaction of humans and the environment;
- intangible evidence, or associative values, by which the landscape is understood and interpreted through the cultural lens of those who create the cultural landscape.

In a significant majority of cultural landscapes now inscribed on the World Heritage List it is tangible evidence that is argued to demonstrate the interaction of humans and their environments. In the designed and continuing cultural landscapes of Europe and Latin America the 'interaction' of humans and the environment is evidenced in built heritage and highly visible large-scale modification of the environment and/or implied through physical association of the cultural evidence with a landform or environmental zone. In general, the interrelationships between the two are not the focus of the values of the property and it is assumed rather than argued that together these natural and cultural 'elements' constitute a cultural landscape. This is also the case in many relic cultural landscapes and especially those associated with indigenous cultures and early humans in which the cultural landscape is defined by the distribution of rock art or archaeological evidence. In these properties interaction with the environment in the past is inferred from the cultural evidence, but the basis on which the spatial distribution of this evidence can be considered a cultural landscape rather than a site or sites is not clear in the outstanding universal values of these properties. In this sense, while these cultural landscapes have increased the diversity of values and places inscribed on the World Heritage List, many are conservative in the evidence used in support of the values and in the interpretation of that evidence.

The indicators of the values of continuing cultural landscapes of Africa and those associated with traditional people in Asia are, like those of Europe and Latin America, associated with land use, architecture and settlement pattern. However, in these properties interpretation of the tangible evidence within the anthropological framework of the traditional knowledge and practices emphasises the interrelationships of the tangible cultural evidence and the environment as an expression of a particular cultural community and defines the boundaries of the property in relation to the cultural values. These properties clearly demonstrate the potential of cultural landscape as a category of World Heritage property to break down the distinction between natural and cultural values in World Heritage properties, not only by interpretation of the landscape through a non-Western cultural lens but through clearly articulating the interrelationships of people and the environment that define the cultural landscape.

The aim of the World Heritage Committee to reconnect culture and nature within the World Heritage system is arguably most fully realised in the associative cultural landscapes inscribed on the World Heritage List. However, only nine have been inscribed in the 20 years since the introduction of the cultural landscapes category. Rather than the cultural landscape being conceptualised as the 'combined works of nature and man' in the values of these properties, a distinction between the two is submerged within a specific cultural understanding of the environment as a whole or in the interpretation of a landform or features. The inscriptions of the indigenous associative landscapes of Tongariro National Park, Uluru Kata-Tjuta National Park and, more recently, Papahānaumokuākea National Marine Monument exemplify the profound conceptual shift in the World Heritage system represented by the introduction of the cultural landscapes category, that is, the recognition that the idea of the 'natural' environment is a cultural construct and culturally relative. It remains to be seen whether this conceptual shift will extend to the interpretation of physical evidence in 'natural' environments. There are no cultural landscape inscriptions in which human interaction with the environment is visible in modification of 'natural' environments, such as human influence on biodiversity.

The emphasis on built and archaeological heritage and large-scale landscape modification in most World Heritage cultural landscapes reflects not only the regions in which the majority are located but also the continuing emphasis on built and monumental heritage in the World Heritage system in general. This offers a partial explanation for why there have been so few inscriptions of indigenous cultural landscapes, other than associative landscapes and those in which rock art provides the primary evidence in support of a cultural landscape inscription. Many indigenous cultural landscapes are in developing countries with scarce resources and capacity for developing World Heritage nominations that may not be successful. In some regions, notably South America, the absence or very small number of indigenous cultural landscape inscriptions is likely to reflect the priorities and interests of national governments. Although there have been some excellent nominations of continuing traditional indigenous and hunter–gatherer landscapes, these are so few in number that they do not provide a well-developed model for nominations of these landscapes, especially where the tangible evidence of the cultural landscape and human interaction with the environment may be ephemeral and/or evident in the values of the 'natural' environment rather than tangible cultural heritage.

NOTE

The UNESCO regions presented here are defined by UNESCO for its activities, and do not necessarily reflect the actual geographical location of countries. All data and description of the outstanding universal values of the World Heritage properties discussed in this chapter are compiled from information available on the World Heritage Centre website, whc.unesco.org.

REFERENCES

Aplin, G. 2007. World Heritage cultural landscapes. *International Journal of Heritage Studies* 13(6): 427–46
Bandarin, F. 2007. Preface, in N. Mitchell, M. Rössler & P–M. Tricaud (ed.) *World Heritage cultural landscapes: a handbook for conservation and management*: 3–4 (World Heritage Papers 26). Paris: UNESCO World Heritage Centre.
Bridgewater, P., S. Arico & J. Scott. 2007. Biological diversity and cultural diversity: the heritage of nature and culture through the looking glass of multilateral agreements. *International Journal of Heritage Studies* 13(4): 405–19.
Fowler, P.J. 2002. *World Heritage cultural landscapes 1992–2002* (World Heritage Papers 6). Paris: UNESCO World Heritage Centre.
Gosden, C. & L. Head. 1994. Landscape—a usefully ambiguous concept. *Archaeology in Oceania*. 29(3): 113–16
Lowenthal, D. 1997. Cultural landscapes. *The UNESCO Courier* 50(9): 18–20.
Mussi, M. 2010. Paleo-landscapes and vulnerability in the framework of the World Heritage Convention, in N. Sanz (ed.) *Human evolution: adaptations, dispersals and social developments (HEADS) World Heritage Thematic Programme*: 190–201 (World Heritage Papers 29). Paris: UNESCO.
Natsuko, A. & T. Sirisrisak. 2008. Cultural landscapes in Asia and the Pacific: implications of the World Heritage Convention. *International Journal of Heritage Studies* 14(2): 176–91.
Plachter, H. & M. Rossler. 1995. Cultural landscape: reconnecting culture and nature, in B. von Droste, H. Plachter & M. Rössler (ed.) *Cultural landscapes of universal value. Components of a global strategy*: 15–18. Jena: Fischer.
Titchen, S. 1996. On the construction of 'outstanding universal value': some comments on the implementation of the 1972 UNESCO World Heritage Convention. *Conservation and Management of Archaeological Sites* 1: 235–42.
UNESCO. 1972. Decision concerning the Convention Concerning the Protection of the World Cultural and Natural Heritage. UNESCO General Conference, 17th Session, 16 November 1972. Available at: http://whc.unesco.org/en/conventiontext (accessed 8 January 2012).
UNESCO. 1987. Note on rural landscape and the World Heritage Convention. World Heritage Committee Meeting (11th Session) Paris, 7–11 December 1987. WHC Document SC-87/CONF.005/INF4. Available at: whc.unesco.org/en/sessions/11COM/documents/ (accessed 8 January 2012).
UNESCO. 1992. *Operational guidelines for the implementation of the World Heritage Convention. 1992*. Paris: UNESCO World Heritage Centre.
UNESCO. 1994a. Expert meeting on the "Global Strategy" and thematic studies for a representative World Heritage List. Unpublished report UNESCO, Paris, 20–22

June 1994. Available at: http://whc.unesco.org/archive/global94.htm#debut (accessed 8 January 2012).

UNESCO. 1994b. *Operational guidelines for the implementation of the World Heritage Convention 1994*. Paris: UNESCO World Heritage Centre.

UNESCO. 2011. *Operational guidelines for the implementation of the World Heritage Convention 2011*. Paris: UNESCO World Heritage Centre.

Contributors

Jennifer Atchison is a researcher at the Australian Centre for Cultural Environmental Research. She is interested in the relationships between plants and people in contemporary life, as well as over longer time frames. Her research utilises scientific, qualitative social science and cultural research methods in geography, environmental science and archaeology. She has worked on research projects about Aboriginal interactions with plants and fire, wheat production and climate change in Australia and contemporary management of invasive plants and animals. Her PhD examined Aboriginal use and management of fruit trees and yams in north-western Australia.

Caroline Bird is a Research Associate in the Department of Anthropology and Archaeology at the Western Australian Museum and an independent archaeologist and consultant. She studied Archaeology at Cambridge and the University of Western Australia. She has research interests in stone artefact analysis and landscape archaeology, and has conducted research in Victoria and Western Australia. She is also interested in applied archaeology, particularly in heritage and public education, and has been involved in the development of curriculum and educational resources in archaeology, heritage and Aboriginal Studies, for students at all levels.

Dr Jean-Christophe Castel has been a researcher in the Department of Archaeozoology, Les Musées de Genève since 2003, having completed a PhD on subsistence activities based on material from Combe-Saunière and Cuzoul de Vers at the University of Bordeaux I. His research focuses on the study of faunal remains from the Middle Palaeolithic to Mesolithic in southwestern France and aspects of taphonomy. He has excavated at several caves sites in France and Switzerland.

Jean-Pierre Chadelle is an archaeologist affiliated with the PACEA laboratory of the University of Bordeaux I (France). He has held positions at the National Museum of Prehistory (Les Eyzies) and the Direction of Prehistoric Antiquities of the Aquitaine region, where his work focused on the curation of sites and decorated caves, such as Lascaux. He has collaborated in the excavation and study of several prehistoric sites in France. With Jean-Michel Geneste, he codirected the excavation of the rich Palaeolithic site of Combe-Saunière. Since 1995, he has headed the preventive archaeology department of the General Council of the Dordogne.

Richard Cosgrove is a Reader in the Archaeology Program at La Trobe University. His research and teaching has focussed on Australian Aboriginal archaeology of the Holocene and late Pleistocene periods. He has worked internationally on projects in France, China, Jordan and England over the last 30 years. He has published the results of his research into Late Pleistocene human behavioural ecology, Aboriginal rock art, the use and impact of anthropogenic fire on various ecosystems, reconstructions of palaeoenvironments, zooarchaeology and hunter–gatherer archaeology.

Peter Davies is a Research Assistant in the Archaeology Program at La Trobe University. He is the author of *Henry's Mill: The Historical Archaeology of a Forest Community* (Archaeopress, 2006) and co-author (with Susan Lawrence) of *An Archaeology of Australia Since 1788* (Springer, 2011). He is co-editor of the journal *Australasian Historical Archaeology* and of the monograph series *Studies in Australasian Historical Archaeology*.

Matthew J. Douglass earned his PhD from the University of Auckland, New Zealand in 2010 and is Research Assistant Professor and Lecturer in the Department of Anthropology and the Center for Great Plains Studies at the University of Nebraska-Lincoln. His research addresses hunter–gatherer ecodynamics and human material culture interaction, most recently in western New South Wales, Australia and the North American Great Plains.

Tim Denham is a Research Fellow in the Archaeology Program at La Trobe University. His research has focused on the emergence and transformation of agriculture in the highlands of Papua New Guinea. In recent years, his spheres of interest have broadened to include: global perspectives on early agriculture, the Holocene histories (focussing on plant exploitation) of Island Southeast Asia, New Guinea and northern Australia, and the contribution of archaeology to understanding current environmental problems. He was the lead author and organiser of Papua New Guinea's successful nomination of the Kuk Early Agricultural Site to UNESCO's World Heritage List in 2008.

Steven E. Falconer is Professor and Chair of Archaeology at La Trobe University. He studies the rise and collapse of urbanised societies in the Eastern Mediterranean and Near East, turning particular attention to the interactions of small agrarian villages with their larger social, political and natural environments. He utilises settlement pattern, ceramic, faunal and metallurgical data to characterise rural life during the periodic development and abandonment of the region's earliest polities and their regional economies. He has directed excavations at Bronze Age Tell el-Hayyat, Tell Abu en-Ni'aj, Dhahret Umm al-Marar and Zahrat adh-Dhra', Jordan and Politiko-*Troullia*, Cyprus.

Patricia L. Fall is Charles La Trobe Professorial Fellow of Geography at La Trobe University. She studies past human and environmental interactions in a variety of geographic settings, including the arid and semi-arid environments of the eastern Mediterranean basin and tropical forests on Pacific and Caribbean islands. She specialises in the analysis of pollen and plant macrofossils from small lakes and bogs, as well as carbonised seeds and wood from archaeological deposits, to reconstruct the interactions of ancient societies with their environments. Her recent research has focused on modelling the present and past landscapes of the eastern Mediterranean.

Associate Professor **Patricia (Trish) C. Fanning** is a geomorphologist based in the Graduate School of the Environment at Macquarie University, with research interests in geoarchaeology, landscape evolution, environmental change and human–environment interaction. She first observed impacts of geomorphic processes on the preservation of the traces of prior Aboriginal occupation whilst undertaking research for her MSc degree. Her PhD research focused on developing a new geo-archaeological approach to understanding human–environment interactions in the past. She has conducted research in western NSW and Cape York Peninsula, Australia, and in Tierra del Fuego in southern Argentina.

Kathryn E. Fitzsimmons is a geochronologist and Quaternary geomorphologist, specialising in optically stimulated luminescence dating. Her research interests focus on the reconstruction of records of human–environmental interaction and environmental change over the Quaternary period, with a particular focus on arid and aeolian landscapes. She investigates human responses to long-term landscape and climate change in the semi-arid regions of Australia, the loess belt of eastern Europe, and Africa, working to produce systematic and high resolution chronologic frameworks. She is presently leader of the luminescence dating laboratory at the Max Planck Institute of Evolutionary Anthropology in Leipzig, Germany.

David Frankel is Professor of Archaeology at La Trobe University. He studied archaeology at the University of Sydney and Gothenburg University, where he specialised in Cypriot prehistory. After some years in the Department of Western Asiatic Antiquities, the British Museum, he returned to Australia in 1978. His research interests include Australian Aboriginal archaeology with particular reference to south-eastern Australia and the archaeology of the Bronze Age in Cyprus, where he has carried out excavations at several sites. He is joint editor-in-chief of *Studies in Mediterranean Archaeology* and a fellow of the Australian Academy of the Humanities.

Professor **Jean-Michel Geneste** is from the Université de Bordeaux 1 and for more than 25 years has carried out research on Palaeolithic sites in France. He manages the primary national rock art research laboratory in France—the Centre National de Préhistoire—the major focus of which is the study of rock art recording, mapping, related archaeological deposits and site management. He has coordinated numerous large archaeological and multidisciplinary research programs in France, Ukraine, South Africa, Papua New Guinea and Australia, including the Chauvet Cave International Research Program.

Professor **Lesley Head** is an Australian Laureate Fellow in the School of Environmental Sciences at the University of Wollongong. Her research focuses on the relationships between people and environments, both conceptual and material. In recent years she has worked in cultural geography, developing an interest in Australian Aboriginal land use, ethnobotany and fire, using palaeoecology and archaeology to study long-term changes in the Australian landscape. She is also involved in interdisciplinary studies of sustainability and climate change. Her publications include *Second Nature: The History and Implications of Australia as Aboriginal Landscape* and *Cultural Landscapes and Environmental Change*.

Professor **Simon J. Holdaway** is an archaeologist in the Department of Anthropology, the University of Auckland, New Zealand. He has research interests in human–environment interactions, geoarchaeology, landscape archaeology and material culture, particularly stone artefacts. He has undertaken research in southwest Tasmania, western New South Wales and Cape York in Australia, the Fayum region of Egypt and in Taranaki, and Great Mercury Island in New Zealand.

Paul Kajewski works in the Archaeology Program at La Trobe University. He completed his first degrees in archaeology at the University of Queensland and subsequently worked for some years as a professional archaeologist in Britain, undertaking fieldwork at a range of different sites.

Susan Lawrence is Associate Professor in the Archaeology Program at La Trobe University. She is interested in the archaeology of British colonisation, gender and material culture studies. Her research has focused on the Australian gold rush, the colonial whaling industry, the pastoral frontier and comparative studies of nineteenth-century British colonies.

She is Past-President of the Australasian Society for Historical Archaeology, serves on the Archaeology Advisory Committee of the Heritage Council of Victoria, is on several editorial boards and is a fellow of the Australian Academy of the Humanities and the Society of Antiquaries of London.

Anita Smith is a Senior Lecturer in the Archaeology Program at La Trobe University. She specialises in the archaeology of the Pacific region with a particular focus on Tahiti and Polynesian voyaging. She has major interests in Heritage both on a global scale and more specifically the relationship between UNESCO cultural heritage programs and decolonisation in the Pacific islands. Anita was a member of Australia's delegation to the World Heritage Committee 2007–2011 and is a member of the Heritage Council of Victoria.

Nicola Stern is a Senior Lecturer in the Archaeology Program at La Trobe University. She has a longstanding interest in landscape archaeology and the way archaeological and geological data are integrated to document the empirical characteristics of landscape scale palimpsests. She has worked on the Lower Pleistocene record of East Africa and is currently running an interdisciplinary project in the Willandra Lakes Region World Heritage Area to investigate the interplay between individual activity traces and the formation of the landscape palimpsest, and the economic, technological and social strategies people developed in response to dramatic and long term environmental changes.

David C. Thomas recently completed his PhD, entitled "The ebb and flow of an empire—Afghanistan and neighbouring lands in the twelfth and thirteenth centuries," at La Trobe University, where he is now an Honorary Research Associate. His first degree was in Archaeology and Anthropology at the University of Cambridge. He completed an MSc in Computing and Archaeology at the University of Southampton in 1994, and has worked in the British Institute in Amman for Archaeology and History, and as a researcher on the Kilise Tepe and Abu Salabikh projects in Cambridge. He currently works in contract archaeology in Victoria.

Jacqueline Tumney is a recent PhD graduate of the Archaeology Program at La Trobe University. She has an honours degree in Science (Earth Sciences) from the University of Melbourne and completed a Master of Archaeology at La Trobe University. Her previous research projects include optical dating of megafauna-bearing sediments in western Victoria and an investigation of archaeological evidence for modern human behaviour in Australia and Africa. Her doctoral research at Lake Mungo, in southwest New South Wales, brought together her interests in geology, Australian Indigenous archaeology, stone tool technology and human behavioural ecology.

Jennifer M. Webb is a Research Fellow in the Archaeology Program at La Trobe University. She specialises in the archaeology of Bronze Age Cyprus and has codirected Australian excavations at Marki, Deneia and Politiko in Cyprus and published numerous collections of Cypriot antiquities in Australia and elsewhere. She is a member of the Editorial Board of the *Journal of Mediterranean Archaeology*, joint editor-in-chief of the monograph series *Studies in Mediterranean Archaeology* and a fellow of the Australian Academy of the Humanities and the Society of Antiquaries of London.

Index

A

Aboriginal culture, people 8, 26, 55–64, 152, 167, 168, 171–2, 176; burning 79; Dreaming 4; language groups 80
adaptation 3–8, 15, 22, 58–9, 69, 85, 111, 138, 145, 149, 150–7, 169
aeolian 42, 52
Afghanistan 6, 84–94
Africa 15, 181–202
agrarian communities 123, 131; economy 7; ruralism 124; societies 7, 123, 127, 131
agriculture 3, 7, 102–3, 152, 154, 167–8: development of 3; hoe-based 7, 135, 140–1
alluvial fan 42–3, 110
Alyawara 56
Anatolia 136
Annales School 6, 84–5
antelope, saiga 16
antler 17, 26; tools 22–3
aquatic resources 42–3
arboriculture 7, 103, 129–31
archaeobotany 105
archaeology: economic 2; environmental 2; historical 149–60; landscape 149–50
arid 24; areas 51, 56; conditions 35; core 31; landscape 91; plains 33; semi- 33
art rock 6, 69, 71–2, 76–8, 184, 186–8, 191, 196–7, 200–1; painted 69; parietal 18; portable 18; stencil 69
Asia central 6, 88; mainland 114
Aurignacian 15, 21–3
Australia 3–8, 24, 31–3, 51–64, 69, 80, 149–60, 167–77, 197, 199

Australian: Alps 31; archaeological record 14; hunter–gatherer 3; research 3, 4
awl 17, 25
Azillian 26

B

Ballarat 152, 154
banana 103, 105, 107–11
baskets 25
Bass land bridge 26
behavioural ecology 58–9
Bendigo 152, 154–5, 158, 160
Bennet's Wallaby 15, 25
Billimina 69–80
Binford, L.R. 51–2, 57–8
birds 53–4, 126
Birmingham, J. 150
bone 15, 17, 19, 22, 26–7, 36, 42, 127–8, 139, 165; assemblages 126, 128; burnt 23; discard 25; long 23, 25; meat-bearing 23; point 25–6, spears 25, technology 26; tools 22, treatment 23
Bone Cave 25–6
bovid 16, 22, 127
blades 21, 25, 71; backed 59; bladelets 21; waisted 108; metal 116
blanks 17, 61
Bowler, J. 35–7, 42
Braudel, F. 85, 91
broad-spectrum foraging 108; diets 109
Bronze Age 9, 17, 123, 129, 135–6
burial: extramural 140, 143; jar 144; multiple 142; pit 142; practices 135
burins 21, 24

burning 55, 79, 108–9, 127, 129–30; anthropic 104, 108
butchery 17–18; selective 23; strategies 19

C

California 151, 156–8
Canaanite 124
carbohydrate 25, 53, 55, 170
cassowaries 115
Castlemaine 155–6
cattle, cattle breeding 126–8, 136, 139, 141, 143
cattle/plough complex 7, 135, 139–40, 170
cave sites 14, 25–6
cereal 90, 126, 128–30, 139–40
ceramic 3, 125, 127, 130–1, 136, 140, 143, 207; ecology 3
Chalcolithic 7, 9, 124, 135–46
chamois 16, 22
charcoal 77, 105, 126–30
Chatelperronian 15–23
chert 25, 72, 74
chiefdoms, chieftains, chiefs 14, 86–7
Chinese 154, 157–8
city-states 14
colonisation, colonists 4, 7, 14, 108, 136, 149–50, 157; re-colonisation 14
collector 51–2, 57–8; *see also* forager
Combe Sauniére 15–26
complex societies 14, 123, 131
contextual approach 7, 135
continental shelf 26, 156
copper 125, 136–7, 140, 142, 158, 191
cores 36, 45, 61, 74; deep sea 60
Cornwall 158, 185, 190
cortex 61–3
Coutts, P.J. 69
Creswick 158–9, 160
crucibles 130–1, 142
culling rates 54
cultivation 7, 101–5, 107–8, 126, 128–30; dryland 111, 114; shifting 109–11; sweet potato 114–16; wetland 111–12
Cyprus 7, 123–32, 135–46

D

Darling River 43
Darwin glass 25
deer 22, 139; fallow 126, 128, 130; hunting 128; red 16
Devon 158, 185, 190
Devon Downs 76

Dioscurea see yam
discard patterns, discard rates 14, 18, 20, 25, 74–5
ditch networks 112–13
Djam 86–94
dogs 115
domestication 55, 103, 117, 168
donkey 136
Dordogne 15
drought 54, 56, 60, 112, 116, 149, 154, 157–8

E

El Niño 54, 112, 116
empires 6, 84
emu: egg 36; eggshell 24; track motif 77
erosion 33, 36–7, 44, 110, 154, 156, 160; blowouts 44; rates 104, 110
ethnography 51, 55–9, 110, 113, 115, 127, 168
Euphrates 129
Europe 14, 18, 23–4, 26, 90, 181–2, 184–5, 195–7, 200

F

faience 143
farmers, farming 6, 102, 123–4, 127, 129–31, 152, 154
fats, fat depletion 25–6
faunal analysis 19, 22–3
fire 53, 55, 110
fish, fishing 43, 127, 176; otoliths 43
flood 149, 157, 160; pulses 33, 42–3, 47
food: foodways 142; -getting strategies 47; plants for 101, 103, 110; production 14, 102–3, 113; remains 36, 44, 47
forage, forager 42, 46–7, 51–2, 58, 102
foraging 43; broad-spectrum 108; expeditions 60; strategies 45–6, 56–8, 61, 108–10
Forde, C.D. 1
forest 7, 105, 108–11, 131; montane 103, 108–10; tropical 102
fossils: trace 33
Fox, C. 1
France 5, 13–15, 23–4, 26, 185
freshwater 35, 42; lake 47; mussels 43
fuel consumption 126–30
furnace 130

G

Gamble, C. 13, 18
Gariwerd 5, 69–82

Ghūrid 6, 84–94
glacial/inter glacial cycles 5, 13
goanna 56
gold 153–160; gold rush 8, 149, 151
Golgol 35, 37–8
Golson, J. 107, 113
Grampians *see* Gariwerd
grassland, grasses 24, 26, 53, 55, 78, 103, 105, 108–11
Gravettian 15–23
Great Dividing Range 152–3, 155
Greece 136
greenstone 72
grindstones 36, 40
ground stone 140
groundwater 35, 42, 47
Gunn, R.G. 76–80

H
Hägerstrand, T. 105
hare 16
Hawkes, K. 56
hearth 36, 41–4, 47, 60, 108, 140, 142; containing terrestrial and/or lacustrine resources 43; fishbone 43; hearthstones 36; radiocarbon ages 62
herbivore 8, 53–4, 171
herding 7, 127–9, 193
Hiscock, P. 59
Holocene 39, 42, 54, 59, 70, 74, 77–9, 80–1, 101, 104, 109–12, 115–64
horse 15–16, 22, 86, 92, 153, 158
horticulture 101, 103, 107, 117; practices 6, 116–17
Hoskins, W.G. 2
Hunter-gatherer, hunting, hunting and gathering 3–5, 13–27, 51–2, 57–8, 69, 102, 109, 115, 127, 129, 167–8, 176, 196–8, 201; *see also* forager; foraging
husbandry, animal 7, 114–15, 127–8, 143
hydraulic sluicing, sluicing 153–4, 157, 159–60
hydrology 31, 33, 36, 156, 160; lake 33, 36, 47

I
Iberia, Iberian Peninsula 18
ibex 17; incisors 17
Ice Age 15, 18; settlement 15; *see also* Pleistocene

India, Indian sub–continent 84, 86, 88, 91–3, 196–7
industrialisation 150
Iron Age 125
irrigation canals 91; facilities 154; technology 157–8
Iran, Iranian world 84, 88, 90
Inter-Tropical Convergence Zone 60
Islamic 86–94
Isle River valley 15
ivory 17, 25

J
Jananginj Njaui 69, 75–6, 79
Jewish trading colonies 93
Jones, R. 14, 172
Jordan 7, 123; Rift 123–4, 127–8, 131; River 124, 127; Valley 130

K
Kalahari 51
kangaroo 54, 59, 171
Karoo 52
Kealhofer, L. 150–1
Keep River 169, 172–5
kiln 125, 131
Kimberley 169, 171
Kirch, P. 140, 144
Kissonerga 137–45
Kuk 104–5, 107–14
Kutikina 25–6

L
Lachlan River 33
lacustrine resources 40, 43, 47
Lake Leaghur 34
lake levels 35, 40–4, 78–9
Lake Mungo 31, 34, 43, 46; *see also* Mungo lunette
Lake Wartook 71–2, 81
landscape: active 37; anthropic 103; anthropogenic 7, 130–1; archaeology 149, 150; archipelagic 91; cognitive approach to 2; cultural 1, 8, 151, 154–5, 169, 181–202; degradation 7, 116; learning 8, 149–60; social approach to 2; structured 101; transported 136; wooded 7
Last Glacial Maximum 9, 13, 18, 21, 24–6, 43, 71, 78–80, 108
Latinis D.K. 102
Laugerie Haute 17

legumes 126, 128–9
Lemba 137–40
Levant 123–4, 126–7
limestone 15, 130–1
lithic 17, 20; debris 19; discard 21;
    material 22; patterns 22;
    see also stone tools
longue durée 18, 85
loom 140
lunette 33–9, 42–7; pelletal-clay 35;
    quartz-sand 35; sandy 43

M

Magdalenian 15, 19–20
mammal 22, 53, 126
mammoth 17
Mallee 80; dune fields 33, 80
Manja 69, 72, 75, 79–80
Marki, Marki-*Alonia* 123–4, 137, 139
marrow 23, 25
Mediterranean 9, 123, 127, 132;
    climate 153; coast 17
meltwater 31, 33
mentalité 2
Mesolithic 26
metallurgy, metallurgical, metal working
    7, 129–31, 136–7, 142–3, 151
migrant 7, 135, 138, 151, 155, 157
migration 86, 136, 145, 149, 151;
    routes 26; zones 54
mineral resources 130–1, 149
mining, miners 149–60
mixed cropping 103
mobile, mobility 45–7, 57, 62–3, 86, 110
Mongol 84–94
Monsoon 60
Morton S.R. 51–5, 59–60, 63
mould 130–1, 140
Mt Alexander 158
Mount Talbot 75
Mousterian 15–23
Mugadgadjin 69, 75, 80
Mungo *see* Lake Mungo, Mungo lunette
Mungo lunette 35, 37–8, 42, 44, 47
Murray-Darling Basin 31, 43
Murray River 81, 155
mussels 43

N

Near East 123
New Guinea *see* Papua New Guinea
New South Wales 5, 9, 51, 59–61, 63,
    76, 151
New Zealand 60, 158, 197, 199
nomads, nomadic peoples 86, 91–2, 198

Nunamira 25–6
Nunamuit 51–2, 57

O

ochre 36, 77
O'Connell, J.F. 56
optimal foraging strategies, theory 56–8
orchards 129–30; *see also* aboriculture
ore 130–1, 136, 142, 157
ornament 17, 22, 136, 143
oven 140, 142
Ovgos Valley 136

P

Pacific 115
palaeoenvironment,
    palaeoenvironmental
    record 5, 23–4, 31–3,
    36–7, 42–7, 59, 198
Palaeolithic 14–23, 24
palimpsest 36, 151, 185, 197
Papua New Guinea, New Guinea 3, 4,
    6, 9, 24, 101–17, 167, 197–8
Pardoe, C. 81
pastoralism 6, 131, 149, 152
pendants 143; *see also* ornaments
Périgueux 15
permafrost 24
Philia 135–46
pigs 108, 112, 114–5, 127–8, 139
plants 55–6, 167–9
Peltenburg, E. 144
Pleistocene 4, 5, 33, 13, 14, 31, 73,
    78–9, 104, 108–10, 115
plough 140; agriculture 136;
    technology 127; zone 137
pollen 24, 108, 114
Politiko-*Troullia* 123–32
Polynesian 115
population shifts, expansion 18, 55
possibilism 1
pottery 7, 124, 127, 130–1, 136–7, 143
Practice–centred method 102–3, 116
prey 15–18, 22, 25, 26; human 22, 25;
    selection 19; terrestrial 47
puddling 153, 155–6, 158, 160
Pyrenees 18, 26, 193
pyrotechnology 123, 126–7, 129–31

Q

quartzite 34, 45–6, 72–3, 196

R

rainforest 108, 110–11, 170, 172, 176;
    montane 103

raw materials; access to 137;
    procurement 6, 81; relative
    importance of sources of 47;
    stone 17, 44–5, 81; distribution
    25
red deer *see* deer
refugia 18, 55
reindeer 15, 16, 17, 22–6
resources *see* aquatic resources;
    lacustrine resources; mineral
    resources; terrestrial resources
riparian 109
risk 116
riverine plain 33
rock art 6, 69, 71–2, 76–80; *see also*
    art; stone carvings
rock shelter, rockshelters 15, 69, 104,
    188, 196–7
rodents 53–4

S
sacred places 112
Sahara 52
Sahel 52
Sahul 24, 26, 108–9
saiga antelope *see* antelope
satellite site 17
satellite images 84, 88
Sauer, C. 1
savannah 24
scale 3, 4, 13, 14, 23–4, 31–3, 55, 75,
    86, 138
scales temporal 3, 13, 55, 57–9, 75, 86;
    chronological 13; of comparison
    23; of observation 5, 59; spatial
    3, 13, 52, 54, 58, 75, 125
scrapers 21, 24, 61, 73
sea level 23, 26, 124
seasonality 17, 23, 25–26, 57–9, 62,
    108, 153, 168–9, 174, 177
secondary products 26, 123, 128, 131
seed 56, 103, 126–30, 171; grinding
    59, 80
sheep 90, 126–8, 136, 139, 152
sheep/goat, ovicaprid 127–8
sheetwash deposits 42
shell, shellfish 43, 168; marine 17, 115;
    middens 43, tools 40
shelters 19, 69, 73, 75–7, 79, 81
shrublands 53
sickle blades 140
silcrete 34, 36, 44–6, 60, 72–5, 81;
    cores 45
slag 130–1
slash-and-burn, slash-and-mulch 109

sluicing *see* hydraulic sluicing
social connections 17; networks 47,
    81, 115
Soffer, O. 13
Solutrean 15–24
South Australia 54, 76
spatulas 26
spear 76; armatures 17; ivory-tipped
    25; thrower hook 17; wooden 25
Stafford Smith D.M 51–2
stone carvings 93
stone tools 15, 19–22, 36, 43–5, 47,
    60–2, 72–6, 111; technological
    study 15
subsistence-level economy 91
sugarcane 111
Sunda 108
sweet potato 108, 114–15
swidden 110; cultivation 109
Syria 129

T
Tambul 113
taro 103, 109–11, 115
Tasmania 5, 13–15, 18–19, 23–6
technology 24–5, 74, 101–2
temperature 18, 24–5, 33, 60, 90, 150
temple 90, 125, 127–8, 131, 188
terrace, terracing 86, 125, 127, 173,
    188, 192, 197–8
terrestrial resources 43, 47
textile 88; production 140
Troodos 125, 129, 136
trade 88, 115, 136, 192, 195
tundra-steppe 24
Turk, mamluks 92–4

U
ungulate 22–3
UNESCO 181–202
Upper Wahgi Valley *see* Wahgi Valley
urban, urbanism, urbanisation 123–4,
    131, 124–5, 128, 131, 150, 154;
    centres 84, 86, 88, 91, 94

V
Vasilia 136
Victoria 69–81, 149–60

W
Wahgi valley 6, 101–17
Walsh, F.J. 55
Warrawau 104, 111, 113
Warreen 25–6
wallaby 15, 26, 54; skin 25–6

Walls of China 37
water 58, 60, 62, 71, 80–1, 86, 153–4;
    channels 91; diversion and
    storage 155–6; management 8,
    149–60; scarcity 8, 149; *see also*
    groundwater; meltwater
weeds 126, 128
Western Desert 56
wetland 43, 81, 104, 109–10;
    cultivation 110–13
whorl 140, 143

Willandra, Willandra Lakes 31–47
wombat 15
woodlands 53, 78, 127, 130
World Heritage Committee 181, 184,
    199, 201; Convention 181–202;
    Register 8, 31, 155,
    181–202
World Systems Theory 85–6

Y
yam 8, 103, 109–10, 115, 167–77